Plant Embryogenesis

METHODS IN MOLECULAR BIOLOGY™

John M. Walker, SERIES EDITOR

436. **Avian Influenza Virus**, edited by *Erica Spackman,* *2008*

435. **Chromosomal Mutagenesis**, edited by *Greg Davis and Kevin J. Kayser, 2008*

434. **Gene Therapy Protocols: Volume 2:** *Design and Characterization of Gene Transfer Vectors* edited by *Joseph M. LeDoux, 2008*

433. **Gene Therapy Protocols: Volume 1:** *Production and In Vivo Applications of Gene Transfer Vectors,* edited by *Joseph M. LeDoux, 2008*

432. **Organelle Proteomics**, edited by *Delphine Pflieger and Jean Rossier, 2008*

431. **Bacterial Pathogenesis:** *Methods and Protocols,* edited by *Frank DeLeo and Michael Otto, 2008*

430. **Hematopoietic Stem Cell Protocols**, edited by *Kevin D. Bunting, 2008*

429. **Molecular Beacons:** *Signalling Nucleic Acid Probes, Methods and Protocols,* edited by *Andreas Marx and Oliver Seitz, 2008*

428. **Clinical Proteomics:** *Methods and Protocols,* edited by *Antonio Vlahou, 2008*

427. **Plant Embryogenesis**, edited by *María F. Suárez and Peter Bozhkov, 2008*

426. **Structural Proteomics:** *High-Throughput Methods,* edited by *Bostjan Kobe, Mitchell Guss, and Huber Thomas, 2008*

425. **2D PAGE: Volume 2:** *Applications and Protocols,* edited by *Anton Posch, 2008*

424. **2D PAGE: Volume 1:,** *Sample Preparation and Pre-Fractionation,* edited by *Anton Posch, 2008*

423. **Electroporation Protocols**, edited by *Shulin Li, 2008*

422. **Phylogenomics**, edited by *William J. Murphy, 2008*

421. **Affinity Chromatography:** *Methods and Protocols, Second Edition,* edited by *Michael Zachariou, 2008*

420. **Drosophila:** *Methods and Protocols,* edited by *Christian Dahmann, 2008*

419. **Post-Transcriptional Gene Regulation**, edited by *Jeffrey Wilusz, 2008*

418. **Avidin-Biotin Interactions:** *Methods and Applications,* edited by *Robert J. McMahon, 2008*

417. **Tissue Engineering**, *Second Edition,* edited by *Hannsjörg Hauser and Martin Fussenegger, 2007*

416. **Gene Essentiality:** *Protocols and Bioinformatics,* edited by *Svetlana Gerdes and Andrei L. Osterman, 2008*

415. **Innate Immunity**, edited by *Jonathan Ewbank and Eric Vivier, 2007*

414. **Apoptosis in Cancer:** *Methods and Protocols,* edited by *Gil Mor and Ayesha Alvero, 2008*

413. **Protein Structure Prediction**, *Second Edition,* edited by *Mohammed Zaki and Chris Bystroff, 2008*

412. **Neutrophil Methods and Protocols**, edited by *Mark T. Quinn, Frank R. DeLeo, and Gary M. Bokoch, 2007*

411. **Reporter Genes for Mammalian Systems**, edited by *Don Anson, 2007*

410. **Environmental Genomics**, edited by *Cristofre C. Martin, 2007*

409. **Immunoinformatics:** *Predicting Immunogenicity In Silico,* edited by *Darren R. Flower, 2007*

408. **Gene Function Analysis**, edited by *Michael Ochs, 2007*

407. **Stem Cell Assays**, edited by *Vemuri C. Mohan, 2007*

406. **Plant Bioinformatics:** *Methods and Protocols,* edited by *David Edwards, 2007*

405. **Telomerase Inhibition:** *Strategies and Protocols,* edited by *Lucy Andrews and Trygve O. Tollefsbol, 2007*

404. **Topics in Biostatistics**, edited by *Walter T. Ambrosius, 2007*

403. **Patch-Clamp Methods and Protocols**, edited by *Peter Molnar and James J. Hickman 2007*

402. **PCR Primer Design**, edited by *Anton Yuryev, 2007*

401. **Neuroinformatics**, edited by *Chiquito J. Crasto, 2007*

400. **Methods in Membrane Lipids**, edited by *Alex Dopico, 2007*

399. **Neuroprotection Methods and Protocols**, edited by *Tiziana Borsello, 2007*

398. **Lipid Rafts**, edited by *Thomas J. McIntosh, 2007*

397. **Hedgehog Signaling Protocols**, edited by *Jamila I. Horabin, 2007*

396. **Comparative Genomics, Volume 2**, edited by *Nicholas H. Bergman, 2007*

395. **Comparative Genomics, Volume 1**, edited by *Nicholas H. Bergman, 2007*

394. **Salmonella:** *Methods and Protocols,* edited by *Heide Schatten and Abraham Eisenstark, 2007*

393. **Plant Secondary Metabolites**, edited by *Harinder P. S. Makkar, P. Siddhuraju, and Klaus Becker, 2007*

392. **Molecular Motors:** *Methods and Protocols,* edited by *Ann O. Sperry, 2007*

391. **MRSA Protocols**, edited by *Yinduo Ji, 2007*

390. **Protein Targeting Protocols** *Second Edition,* edited by *Mark van der Giezen, 2007*

389. **Pichia Protocols**, *Second Edition,* edited by *James M. Cregg, 2007*

388. **Baculovirus and Insect Cell Expression Protocols**, *Second Edition,* edited by *David W. Murhammer, 2007*

387. **Serial Analysis of Gene Expression (SAGE):** *Digital Gene Expression Profiling,* edited by *Kare Lehmann Nielsen, 2007*

386. **Peptide Characterization and Application Protocols,** *edited by Gregg B. Fields, 2007*

385. **Microchip-Based Assay Systems:** *Methods and Applications,* edited by *Pierre N. Floriano, 2007*

384. **Capillary Electrophoresis:** *Methods and Protocols,* edited by *Philippe Schmitt-Kopplin, 2007*

383. **Cancer Genomics and Proteomics:** *Methods and Protocols,* edited by *Paul B. Fisher, 2007*

382. **Microarrays, Second Edition:** *Volume 2, Applications and Data Analysis,* edited by *Jang B. Rampal, 2007*

381. **Microarrays, Second Edition:** *Volume 1, Synthesis Methods,* edited by *Jang B. Rampal, 2007*

METHODS IN MOLECULAR BIOLOGY™

Plant Embryogenesis

Edited by

María F. Suárez

Departamento de Biología Molecular y Bioquímica, Facultad de Ciencias, Universidad de Málaga, Málaga, Spain

and

Peter V. Bozhkov

Department of Plant Biology and Forest Genetics, Swedish University of Agricultural Sciences, Uppsala, Sweden

Humana Press

Editors
María F. Suárez
Departamento de Biología Molecular y
  Bioquímica
Facultad de Ciencias
Universidad de Málaga, Málaga, Spain
fsuarez@uma.es

Peter V. Bozhkov
Department of Plant Biology and Forest
  Genetics,
Swedish University of Agricultural Sciences
Uppsala, Sweden
Peter.Bozhkov@vbsg.slu.se

Series Editor
John M. Walker, Professor Emeritus
School of Life Sciences,
University of Hertfordshire
Hatfield
Hertfordshire AL10 9AB, UK

ISBN: 978-1-61737-848-5   e-ISBN: 978-1-59745-273-1

Cover illustration: From Fig. 2 in Chapter 8, "Isolation of Embryo-Specific Mutants in *Arabidopsis:* Genetic and Phenotypic Analysis", by Nai-You Liu, Zhi-Feng Zhang, and Wei-Cai Yang.

Printed on acid-free paper

9 8 7 6 5 4 3 2 1

springer.com

Preface

Plant embryology is inherently fascinating, for it holds the promise of understanding how a relatively simple and phylogenetically conserved embryonic pattern is transformed into whimsical diversity of growth forms in the flora around us. For example, a colossal pine tree and an annual *Arabidopsis thaliana* develop from the embryos, which have passed through a very similar and stereotyped sequence of stages. What governs a zygote of pine, *A. thaliana* or any other plant species to divide precisely and behave predictably so as to start a new organism with its distinctive and familiar developmental pathway? Finding an answer to this intriguing question – the driving force of modern plant embryology – requires a powerful arsenal of methods to track, control and analyze embryogenesis.

The methods to study plant embryogenesis have undergone a rapid evolution over the past decade. With the advent of postgenome era in biology, it has become possible to generate embryo-specific mutants by T-DNA insertional mutagenesis, and more recently by RNA interference, to use fluorescent protein probes for live cell imaging in the embryos, and to analyze gene expression in few-celled embryonic domains. These and other impressive advances in development of new tools in plant embryology have inclined us to compile the scattered information on the methods used in the different laboratories for understanding plant embryo development and bring them to one platform.

While preparing this volume of *Methods in Molecular Biology*, our idea has been to focus on a common developmental process, so as to bring together the methods developed specifically for studying plant embryogenesis. For this reason, the book does not include those methods, which are applicable to almost every developmental system and can thus be found elsewhere. Although this book is aimed primarily at those who are relatively new to the field, we are sure that those who are fully fledged plant embryologists will also find it useful. We would particularly like to hear from scientists who have had suggestions for short-cuts or modifications to the protocols listed. We thank all our colleagues who made suggestions and offered advice. Finally, we are indebted to all of the authors who contributed to the book and to John M. Walker of Humana Press who oversaw the project and dealt with much of the layout.

María F. Suárez, PhD
Peter V. Bozhkov, PhD
Malaga, Spain and Uppsala, Sweden
March, 2007

Contents

Preface .... *v*

Contributors .... *ix*

PART I: MODEL EMBRYONIC SYSTEMS
   1. Arabidopsis Embryogenesis
      Soomin Park and John J. Harada *3*

   2. Maize Embryogenesis
      Pilar Fontanet and Carlos M. Vicient *17*

   3. Spruce Embryogenesis
      Sara von Arnold and David Clapham *31*

PART II: CELLULAR, GENETIC AND MOLECULAR MECHANISMS
OF PLANT EMBRYOGENESIS
   4. In Vitro Fertilization With Isolated Higher Plant Gametes
      Erhard Kranz, Yoichiro Hoshino, and Takashi Okamoto *51*

   5. In Vitro Culture of *Arabidopsis* Embryos
      Michael Sauer and Jiří Friml *71*

   6. Culture and Time-Lapse Tracking of Barley
      Microspore-Derived Embryos
      Simone de F. Maraschin, Sandra van Bergen, Marco Vennik,
      and Mei Wang *77*

   7. Isolation of Embryo-Specific Mutants in *Arabidopsis*: *Plant*
      Transformation
      Nai-You Liu, Zhi-Feng Zhang, and Wei-Cai Yang *91*

   8. Isolation of Embryo-Specific Mutants in *Arabidopsis*:
      Genetic and Phenotypic Analysis
      Nai-You Liu, Zhi-Feng Zhang, and Wei-Cai Yang *101*

9. Laser-Capture Microdissection to Study Global
    Transcriptional Changes During Plant Embryogenesis
    **Stuart A. Casson, Matthew W. B. Spencer,
    and Keith Lindsey** *111*

10. Promoter Trapping System to Study Embryogenesis
    Robert Blanvillain and Patrick Gallois *121*

11. Visualization of Auxin Gradients in Embryogenesis
    Michael Sauer and Jiří Friml *137*

12. Intercellular Trafficking of Macromolecules
    During Embryogenesis
    Insoon Kim and Patricia C. Zambryski *145*

13. Immunolocalization of Proteins in Somatic Embryos:
    Applications for Studies on the Cytoskeleton
    Andrei P. Smertenko and Patrick J. Hussey *157*

14. Detection of Programmed Cell Death in Plant Embryos
    **Lada H. Filonova, María F. Suárez,
    and Peter V. Bozhkov** *173*

Subject Index *181*

Contributors

ROBERT BLANVILLAIN • *Plant Gene Expression Center, University of California, USA*

PETER V. BOZHKOV • *Department of Plant Biology and Forest Genetics, Swedish University of Agricultural Sciences, Uppsala, Sweden*

STUART A. CASSON • *The Integrative Cell Biology Laboratory, Durham University, UK*

DAVID CLAPHAM • *Department of Plant Biology and Forest Genetics, Swedish University of Agricultural Sciences, Uppsala, Sweden*

LADA H. FILONOVA • *Department of Wood Sciences, Swedish University of Agricultural Sciences, Uppsala, Sweden*

PILAR FONTANET • *Laboratori de Genetica Molecular i Vegetal, Consorci CSIC-IRTA, Spain*

JIŘÍ FRIML • *ZMBP Developmental Genetics, University of Tübingen, Germany*

PATRICK GALLOIS • *Faculty of Life Sciences, University of Manchester, UK*

JOHN J. HARADA • *Section of Plant Biology, College of Biological Sciences, University of California, USA*

YOICHIRO HOSHINO • *Universität Hamburg, Biozentrum Klein Flottbek und Botanischer Garten, Entwicklungsbiologie und Biotechnologie, Ohnhorststr. 18, 22609 Hamburg, Germany*

PATRICK J. HUSSEY • *The Integrative Cell Biology Laboratory, Durham University, UK*

INSOON KIM • *Department of Biology, Sungshin Women's University, Korea*

ERHARD KRANZ • *Universität Hamburg, Biozentrum Klein Flottbek und Botanischer Garten, Entwicklungsbiologie und Biotechnologie, Ohnhorststr. 18, 22609 Hamburg, Germany*

NAI-YOU LIU • *The Key Laboratory of Molecular and Developmental Biology, Institute of Genetics and Developmental Biology, Chinese Academy of Sciences, China*

KEITH LINDSEY • *The Integrative Cell Biology Laboratory, Durham University, UK*

SIMONE DE F. MARASCHIN • *Institute of Biology, Leiden University, Clusius Laboratory, The Netherlands*

TAKASHI OKAMOTO • *Universität Hamburg, Biozentrum Klein Flottbek und Botanischer Garten, Entwicklungsbiologie und Biotechnologie, Ohnhorststr. 18, 22609 Hamburg, Germany*

SOOMIN PARK • *Horticultural Biotechnology Division, National Horticultral Research Institute, Republic of Korea*

MICHAEL SAUER • *ZMBP Developmental Genetics, University of Tübingen, Germany*

ANDREI P. SMERTENKO • *The Integrative Cell Biology Laboratory, Durham University, UK*

MATTHEW W. B. SPENCER • *The Integrative Cell Biology Laboratory, Durham University, UK*

MARÍA F. SUÁREZ • *Departamento de Biología Molecular y Bioquímica, Facultad de Ciencias, Universidad de Málaga, Málaga, Spain*

SANDRA VAN BERGEN • *Fytagoras BV and TNO Quality of Life, The Netherlands*

MARCO VENNIK • *Fytagoras BV and TNO Quality of Life, The Netherlands*

CARLOS M. VICIENT • *Laboratori de Genetica Molecular i Vegetal, Consorci CSIC-IRTA, Spain*

SARA VON ARNOLD • *Department of Plant Biology and Forest Genetics, Swedish University of Agricultural Sciences, Uppsala, Sweden*

MEI WANG • *Institute of Biology, Leiden University, Clusius Laboratory and TNO Quality of Life, The Netherlands*

WEI-CAI YANG • *The Key Laboratory of Molecular and Developmental Biology, Institute of Genetics and Developmental Biology, Chinese Academy of Sciences, China*

PATRICIA C. ZAMBRYSKI • *Department of Plant and Microbial Biology, University of California, USA*

ZHI-FENG ZHANG • *The Key Laboratory of Molecular and Developmental Biology, Institute of Genetics and Developmental Biology, Chinese Academy of Sciences, China*

I

MODEL EMBRYONIC SYSTEMS

1

Arabidopsis Embryogenesis

Soomin Park and John J. Harada

Summary

Embryogenesis in higher plants consists of two major phases, morphogenesis and maturation. Morphogenesis involves the establishment of the embryo's body plan, whereas maturation involves cell expansion and accumulation of storage macromolecules to prepare for desiccation, germination and early seedling growth. *Arabidopsis* mutants showing defects in embryogenesis have provided information for understanding the events that govern embryo formation through molecular, genetic and biochemical analyses. Thus, many of the processes that underlie embryogenesis are beginning to be understood. In this chapter, we focus on genes that play key roles in the morphogenesis phase of *Arabidopsis* embryogenesis.

Key Words: Embryo; maturation; morphogenesis; root apical meristem; seeds; shoot apical meristem.

1. Introduction

Embryogenesis is a crucial developmental period in the life cycle of flowering plants. During embryogenesis, the single-celled zygote follows a defined pattern of cell division and differentiation to form the mature embryo. Development of the embryo occurs in two distinct phases. There is first a phase of morphogenesis during which the basic body plan of the plant is established. This is followed by the maturation phase during which processes related to the embyro's entry into metabolic quiescence and subsequent germination occur. In this review, we discuss the morphogenesis phase of embryogenesis in the model plant *Arabidopsis thaliana* (L.) Heynh. For additional details, readers are referred to other reviews of embryogenesis (**refs. *1–7***).

From: *Methods in Molecular Biology, vol. 427: Plant Embryogenesis*
Edited by: M. F. Suárez and P. V. Bozhkov © Humana Press, Totowa, NJ

2. Morphological Development of the Embryo

Embryogenesis begins with double fertilization in which one sperm cell fuses with the egg cell and another fuses with the central cell to form the zygote and endosperm mother cell, respectively. The zygote then divides asymmetrically to produce an apical cell that develops into an embryo proper and a basal cell that generates the hypophysis and the suspensor (*see* **Fig. 1a**). The hypophysis will give rise to the root quiescent center and the initials of the central root cap, whereas the suspensor is a transient organ that plays structural and physiological roles in embryo development *(8)*.

Morphological development of *Arabidopsis* embryos follows the Onagrad-type of embryogenesis *(9)*. The apical cell undergoes two longitudinal divisions (*see* **Fig. 1b**) followed by a transverse division to form an eight-celled, octant-stage proembryo (*see* **Fig. 1c**). Each cell of the octant-stage embryo proper undergoes a periclinal (parallel to the axis/surface) division thereby generating the protoderm or embryonic epidermis and a dermatogen-stage globular embryo (*see* **Fig. 1d and e**). Localized cell divisions lead to the emergence of cotyledon lobes and a shift in the morphological symmetry of the embryo proper from radial to bilateral with formation of the heart-stage embryo (*see* **Fig. 1f**). By the torpedo stage, the basic embryonic elements of the plant body are in place (*see* **Fig. 1g**) *(2)*. Along the apical–basal axis are the shoot apical meristem, cotyledons, hypocotyl, radicle and the root apical meristem. Along the radial axis, the embryo consists of three primary tissues: the outer protoderm, the middle ground tissues and the inner procambium tissues. Although the basic plant body is formed during embryogenesis, the vast majority of organ and tissue formation occurs postembryonically.

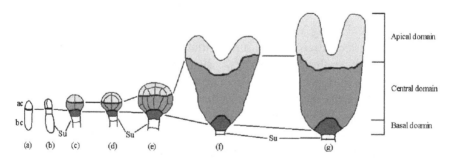

Fig. 1. A schematic drawing showing representative stages of *Arabidopsis* embryo development, with emphasis on the generation of apical, central and basal embryonic domains. (a) Two-celled embryo, (b) four-celled embryo-proper stage embryo, (c) octant-stage embryo, (d) dermatogen-stage embryo, (e) globular-stage embryo, (f) heart-stage embryo, (g) torpedo-stage embryo. ac, apical cell; bc, basal cell; su, suspensor.

Genetic and molecular studies of *Arabidopsis* embryogenesis have identified genes that control specific aspects of embryo development. Major processes in embryo development and the genes involved in these processes are discussed.

3. Asymmetric Division of the Zygote

The first division of the *Arabidopsis* zygote is asymmetric in that the two cells have different fates. This division produces a small cytoplasmically dense apical cell that will become the embryo proper and a larger vacuolated basal cell that gives rise to the hypophyseal cell and suspensor (*see* **Fig. 1a**). The different fates of the two daughter cells are marked by differential gene expression. *ARABIDOPSIS THALIANA MERISTEM LAYER1 (ATML1)*, a class IV homeodomain-leucine zipper gene normally active later in development in the L1 layer of the shoot apex, is expressed in the apical but not basal cell *(10)*. *WUSCHEL-RELATED HOMEOBOX (WOX)* genes are also differentially expressed in apical and basal cells *(11)*. *WOX2* and *WOX8* are expressed in the egg cell and in the zygote, but their expression becomes restricted to the apical cell and basal cell, respectively, after the first division *(11)*.

Apical cell fate correlates with auxin activity in that auxin responsive gene activity is higher in apical than basal cells *(12)*. For example, a synthetic, auxin-inducible promoter, *DR5rev*, drives expression of a reporter gene encoding green fluorescent protein primarily in the apical cell. This differential auxin activity is likely to result from expression of the *PIN FORMED7* (*PIN7*) gene. *PIN7* encodes an auxin efflux facilitator, and the protein is located in the apical membrane of the basal cell following the asymmetric division of the zygote *(12)*.

As will be discussed, auxin transport and the localization of auxin maxima are required for many aspects of embryogenesis and postembryonic development *(13–15)*. Thus, correct expression of the four PIN auxin efflux facilitator genes expressed in embryos, *PIN1*, *PIN3*, *PIN4* and *PIN7*, and proper localization of their gene products is important *(12,16)*. Mutation of a single *PIN* gene does not cause major defects of embryo development, suggesting genetic redundancy. However, the quadruple mutation of *pin1*, *pin3*, *pin4* and *pin7* causes significant defects in apical–basal organization *(12)*. The importance of PIN placement to establish apical–basal polarity is shown by the *gnom* (*gn*) mutant. *gn* mutant embryos appear as a ball-shaped structure without apical–basal polarity, and the plane of the first division of the zygote is often disturbed by the *gn* mutation *(17,18)*. *GN* encodes a guanine nucleotide exchange factor for ADP ribosylation factor involved in vesicular transport, and it is required for proper localization of PIN1 *(19,20)*.

A (MAP) mitogen-activated protein kinase signaling pathway is also implicated in the specification of basal cell lineage. Loss-of-function mutations in
the *YODA* gene encoding a MAPkinase kinase cause defects in elongation of
basal cell derivatives, whereas gain-of-function alleles result in a long suspensor
phenotype and suppression of embryo proper development *(21)*. However, other
components of a MAP kinase signaling pathway involved in apical–basal cell
specification have not been identified.

4. Apical–Basal Patterning

Polarity along the apical–basal axis is observed in the zygote, with the
nucleus and much of the cytosol localized to the apical end and a large vacuole
in the basal end *(9,22)*. By the octant stage, the embryo has become subdi-
vided into three domains upon which morphological structures are formed (*see*
Fig. 1c) *(2)*. The top and bottom halves of the embryo proper corresponds to
the apical and central domains, respectively, whereas the uppermost cell of
basal cell derivatives, the hypophysis, constitutes the basal domain. The shoot
apical meristem and most of the cotyledons derive from the apical domain, part
of the cotyledons, the hypocotyl, radicle and part of the root apical meristem
come from the central domain, and the quiescent center and central root cap
initials of the root apical meristem are from the basal domain.

4.1. The Apical Domain

The apical domain is divided into three subdomains that generate the shoot
apical meristem, the cotyledon boundaries, and the cotyledons. The existence of
the apical embryo domain was suggested initially by discovery of the *GURKE*
gene *(17)*. *gurke* mutant seedlings are defective in cotyledon and shoot apical
meristem formation, although they do possess a hypocotyl, radicle and root
apical meristem *(23)*. Although *GURKE* has been shown to encode an acetyl
CoA carboxylase that catalyzes malonyl CoA synthesis *(24)*, the mechanisms
by which it affects formation of the apical domain are not yet understood.

Studies of the *TOPLESS* gene showed that the fate of the apical domain is
not fixed early in embryogenesis. *topless* mutants lack an apical domain *(25)*.
However, unlike *gurke* mutants, *topless* mutants in some cases develop a root
instead of a shoot, resulting in a seedling with roots at both ends. Because *topless*
is a temperature-sensitive mutation, temperature-shift experiments showed that
apical fate does not become fixed until the transition stage. *TOPLESS* has been
suggested to encode a transcriptional co-repressor that inhibits basal patterning
in the apical domain *(26)*.

The shoot apical meristem consists of a small population of stem cells that
continuously produces cells that become incorporated into the aerial organs of a

plant, the stem, leaf and flower. Studies have identified two classes of mutations affecting postembryonic shoot apical meristem formation. Mutations in *SHOOT MERISTEMLESS (STM)*, *WUSCHEL (WUS) and CUP-SHAPED COTYLEDON (CUC)* result in seedlings with no or reduced shoot apical meristems *(27–29)*, whereas *clavata1 (clv1)*, *clv2* and *clv3* mutants have enlarged shoot apical meristems *(30–32)*.

The shoot apical meristem can first be delineated histologically in a heart-stage embryo with the emergence of cotyledon bulges that flank the meristem *(27)*. However, patterning of the shoot apical meristem begins during the globular stage. The earliest observed event in shoot apical meristem patterning during embryogenesis is the expression of *WUS*, a gene encoding a homeodomain transcription factor, in the four inner apical cells of the dermatogen-stage embryo *(33)*. Studies of postembryonic meristems have shown that *WUS* is expressed in a small region immediately under the stem cells known as the organizing center and is critical for maintenance of the shoot apical meristem by causing adjacent cells to acquire stem cell fate and remain indeterminate *(33)*. During the globular stage of embryogenesis, the *WUS* expression domain becomes further restricted into the organizing center near the center of the embryo *(33)*. The size of the organizing center is regulated through a feedback loop involving the *CLV* and *WUS* genes (reviewed in **refs. *34,35***). WUS stimulates the activity of *CLV3* encoding a secreted polypeptide. Proteins encoded by *CLV1* and *CLV2* form a receptor complex that interacts specifically with CLV3, inducing a signaling pathway that inhibits *WUS* expression and limits size of the organizing center. *CLV* genes are expressed by the heart stage of embryogenesis to maintain meristem size. Another positive regulator of stem cell identity, *STM*, is expressed at the position of the nascent shoot apical meristem beginning at the globular stage *(36)*. STM, a class I KNOTTED-like homeobox transcription factor, acts independently of WUS to prevent differentiation of stem cells *(37)*, primarily by inhibiting ASYMMETRIC LEAVES1, a MYB transcription factor expressed in differentiating cells in lateral organ primordia *(38)*.

Patterning of the apical domain also involves cotyledon formation. *CUC1*, *CUC2* and *CUC3* encoding NAC transcription factors are primarily responsible for forming boundaries between the cotyledons *(29,39–41)*. At the transition stage, *CUC* genes are expressed in a band across the apex, coincident with the expression of *STM*. The *CUC* and *STM* expression domain defines a region that does not contribute to emerging cotyledon primoridia. After the heart stage, *STM* expression becomes restricted to the shoot apical meristem, whereas *CUC* genes show a complementary expression pattern within this band. Thus, *CUC* genes form boundaries that permit the two cotyledons to form and prevent outgrowth of the incipient shoot apical meristem. Two genes have been implicated to play roles in generating cotyledon primordia, *PINOID*, a serine/threonine kinase that controls targeting of PIN1, and *ENHANCER OF PINOID (42)*. Digenic mutations in these genes result in precise deletion of cotyledons with no visible effects on hypocotyls and roots. The double mutation causes a reversal of PIN1 localization in the epidermis. These results suggest that *PINOID* and *ENHANCER OF PINOID* are required to create auxin maxima essential for cotyledon formation.

4.2. Pattern Formation in the Central Domain

The central domain gives rise to the hypocotyl, radicle and part of the cotyledons (*see* **Fig. 1g**). The hypocotyl and radicle constitute the majority of the embryonic axis and can be represented as a cylinder containing concentric layers of tissues *(1)*. Thus, processes that form this axis are important for morphological development within this domain.

The central domain was defined originally by *fackel* (*fk*) mutants that produce seedlings consisting of the shoot apical meristem and cotyledons attached to the root *(17)*. *FK* encodes a sterol C-14 reductase involved in sterol biosynthesis *(43)*, and other mutations that cause Fk-like mutant phenotypes, such as *hydra1* and *cephalopod*, also encode enzymes involved in sterol biosynthesis *(44–47)*. One interpretation of this result is that sterols may act as signaling molecules for pattern formation along the shoot–root axis. Potentially consistent with this hypothesis, *BRASSINOSTEROID-INSENSITIVE1* acts downstream of *FK (48)*. Alternatively, the finding that *fk*, *hydra1* and *cephalopod* mutations cause a deficiency in cellulose and defects in cell wall formation may explain the Fk-like mutant phenotypes *(49)*. Cellulose accumulation may be defective in these mutants because a glucose-conjugated form of a sterol, sitosterol, is thought to serve as a primer for cellulose synthesis *(50)*.

Monopteros (*mp*) mutant seedlings have only a shoot apical meristem and cotyledons and were interpreted to be defective in both the central and basal domains (*see* **Fig. 1g**) *(17,51)*. The earliest defect observed in *mp* mutant embryos is transverse rather than longitudinal divisions of the apical cell of the zygote (*see* **Fig. 1b**), resulting in an octant-stage embryo with four instead of two layers of cells along the axis *(51)*. Later in embryogenesis, the *mp* mutation causes defects in the orientation of cell divisions, such that cells in the central domain fail to form characteristic cell files within the hypocotyl and radicle. This Mp mutant phenotype is also caused by mutations in *BODENLOS* (*BDL*) and *AUXIN RESISTANT6* (*AXR6*), all of which encode proteins involved in auxin signaling *(52,53)*. *MP* encodes an auxin response factor (ARF5), a transcription factor that regulates auxin-responsive genes *(54)*. BDL is an Aux-IAA protein (IAA12) that is thought to bind with MP and inhibit its transcriptional activation function until BDL is degraded in the auxin-response pathway *(55–57)*. *AXR6* encodes a cullin protein of the E3 ubiquitin ligase required for auxin responses *(58)*. Because auxin is likely to be transported in the embryo from the shoot apex through the hypocotyl and radicle to the root apical meristem, these findings suggest that MP, BDL and AXR6 are required to respond to polar auxin transport to establish the embryonic axis.

4.3. Establishment of the Root Apical Meristem

The root apical meristem is responsible for development of the below-ground plant organs and consists of a slowly dividing group of cells, known as the quiescent center surrounded by initial or stem cells that produce the cell files that constitute the root apical meristem. Thus, the quiescent center is analogous to the shoot apical meristem organizing center. The root apical meristem derives from both the central and basal domains (*see* **Fig. 1g**). Initials that give rise to the vascular cylinder and the ground tissue of the root come from derivatives of the central domain, whereas the quiescent center and central root cap initials derive from the hypophysis that corresponds to the basal domain.

At the globular stage, the hypophysis divides asymmetrically to give rise to an apical lens-shaped cell that will become the quiescent center and a basal cell from which the central root cap initials are derived. The timing of this division correlates roughly with a shift in auxin maxima within the embryo from the embryo proper to the hypophysis that appears to be mediated, at least in part, by the PIN1, PIN4 and PIN7 auxin efflux facilitators *(12)*. The finding that *mp* and *bdl* mutant embryos do not form lens-shaped cells of the hypophysis nor functional root apical meristems suggests a requirement for auxin signaling in root apical meristem formation *(51,52)*. Both *MP* and *BDL* are expressed in cells derived from the apical and central domains but not in basal-cell derivatives, suggesting that signaling occurs between cells in the central domain and the hypophysis *(52,54)*.

Response to the auxin signal may be mediated at least in part by the *PLETHORA* (*PLT*) genes, *PLT1* and *PLT2* *(59)*. At the early heart stage, digenic mutations in these genes cause defects in divisions of hypophyseal cell derivatives that later result in misspecification of the quiescent center and surrounding initial cells. *PLT* genes encode putative APETALA2-domain transcription factors whose expression is dependent on auxin and auxin responsive transcription factors, including MP. At the globular stage, these genes are expressed in the lens-shaped hypophyseal cell and provascular cells, but this expression domain later becomes restricted to the quiescent center and surrounding initial cells. The *PLT* genes have been proposed to interact with two other genes required for specification of the quiescent center, *SCARECROW*

(*SCR*) and *SHORTROOT* (*SHR*) *(60)*. When expressed ectopically, *PLT* genes specify a new quiescent center and initial cells in positions in which its expression domain coincides with those of *SCR* and *SHR* *(59)*. Together with *SCR* and *SHR*, the auxin-dependant *PLT* genes form a combinatorial code to specify stem cells, suggesting a role of auxin signal transduction in early embryonic specification of meristem domains.

The *HOBBIT* (*HBT*) gene is required for cell division and cell type specification in the root apical meristem *(61,62)*. At the globular stage, *hbt* mutant embryos either do not form the lens-shaped cell of the hypophysis or undergo atypical divisions in this region. *hbt* mutant seedlings lack columella and lateral root cap cells and a quiescent center. Cell division does not occur in the postembryonic root meristem. Cell fates in the root apical meristem also appears to be affected by the mutation. *HBT* encodes a homolog of the CDC27 subunit of anaphase-promoting complex that is required for cell cycle progression *(63)*.

5. Radial Pattern Formation

A transverse section through the hypocotyl or radicle of the mature embryo reveals the concentric arrangement of protoderm, ground meristem and procambium. Radial patterning refers to processes that give rise to these different outer to inner embryonic tissue systems. The first visual manifestation of a tissue occurs when the octant-stage embryo undergoes periclinal divisions to give rise to an outer layer of protoderm cells and inner cells (*see* **Fig. 1d**). Further morphological discrimination of inner tissues occurs during the globular stage as cells in the center of the central domain undergo periclinal divisions to give rise to the vascular cylinder (*see* **Fig. 1e**).

Specification of protoderm cell fate may occur at the earliest stage of embryogenesis. *ATML1* is expressed specifically in the apical cell following asymmetric division of the zygote and continues to be expressed throughout the embryo proper up to the octant stage *(10)*. *ATML1* expression becomes restricted to the protoderm or L1 layer following periclinal divisions that produce the dermatogen-stage embryo, and it is not detected in inner cells. *ATML1* expression remains L1 specific later in development, although it is expressed primarily in the epidermis of the shoot apex. A similar expression pattern is observed for *PROTODERMAL FACTOR2* (*PDF2*) *(64,65)*. One interpretation of these results is that signals either external to the embryo or imbedded in the cell wall of the zygote may confer protoderm cell fate. However, mutations in two genes involved in cytokinesis, *KNOLLE* and *KEULE* *(66,67)*, cause the expression of epidermis-specific genes in subepidermal tissues and incomplete cell wall formation in embryos *(68)*. Thus, the signals that specify protoderm cell fate remain unclear.

The *wooden leg* (*wol*) mutation has implicated cytokinins in differentiation of vascular tissues. In *wol* mutant embryos after the heart stage, the number of cell divisions in the hypocotyl and vascular cylinder are reduced and phloem is reduced or absent *(61)*. *WOL* is allelic with *CYTOKININ RESPONSE1*, a two-component histidine kinase implicated as a cytokinin receptor *(69,70)*.

The ground meristem initials in the root apical meristem undergo periclinal divisions resulting in the formation of the cortex and endodermis layers. Mutations in *SHR* and *SCR* result in defects in the cortex and endodermis as early as the torpedo stage *(61)*. Both *SHR* and *SCR* encode GRAS-type transcription factors required for asymmetric divisions of the ground tissue initials and endodermis specification *(71–73)*. *SHR* is expressed in the vascular cylinder, but the SHR protein moves to the adjacent ground tissue layer where it activates *SCR*. *SCR* promotes the asymmetric periclinal division in the ground tissue initial and specification of endodermal cells *(60,74)*.

6. Prospects

Great progress has been made during the past 20 years in understanding the processes that underlie embryogenesis. Molecular and genetic approaches have defined genes that control key aspects of embryo development. However, much remains to be done to obtain a comprehensive understanding of embryogenesis. It is likely that many regulators of embryonic processes have not yet been identified because of problems of genetic redundancy. Moreover, much work remains to be done to identify the genes and cellular processes that act downstream of these key embryonic regulators. The framework established for understanding embryogenesis should accelerate future progress in this research area.

Acknowledgments

We thank the members of the Harada lab for their helpful comments about this review. Support from NSF and DOE is acknowledged.

References

1. Berleth T, Chatfield S. Embryogenesis: pattern formation from a single cell. In: Somerville CR, Meyerowitz EM, eds. The Arabidopsis Book. Rockville: American Society of Plant Biologists, 2002:1–22.

2. Jurgens G. Apical-basal pattern formation in Arabidopsis embryogenesis. EMBO J 2001;20:3609–16.

3. Laux T, Wurschum T, Breuninger H. Genetic regulation of embryonic pattern formation. Plant Cell 2004;16:S190–202.

4. Willemsen V, Scheres B. Mechanisms of pattern formation in plant embryogenesis. Annu Rev Genet 2004;38:587–614.

5. Yadegari R, Goldberg RB. Embryogenesis in dicotyledonous plants. In: Larkins BA, Vasil I.K eds. Cellular and Molecular Biology of Plant Seed Development. Dordrecht: Kluwer Academic Publishers, 1997:3–52.

6. Goldberg RB, de Paiva G, Yadegari R. Plant embryogenesis: zygote to seed. Science 1994;266:605–14.

7. West MA, Harada JJ. Embryogenesis in higher plants: an overview. Plant Cell 1993;5:1361–9.

8. Yeung EC, Meinke DW. Embryogenesis in angiosperms: development of the suspensor. Plant Cell 1993;5:1371–81.

9. Mansfield SG, Briarty LG. Early embryogenesis in *Arabidopsis thaliana*. II. The developing embryo. Can J Bot 1990;69:461–76.

10. Lu P, Porat P, Nadeau JA, O'Neill SD. Identification of a meristem L1 layer-specific gene in Arabidopsis that is expressed during embryonic pattern formation and defines a new class of homeobox genes. Plant Cell 1996;8:2155–68.

11. Haecker A, Gross-Hardt R, Geiges B, Sarkar A, Breuninger H, Herrmann M, Laux T. Expression dynamics of *WOX* genes mark cell fate decisions during early embryonic patterning in *Arabidopsis thaliana*. Development 2004;131:657–68.

12. Friml J, Vieten A, Sauer M, Weijers D, Schwarz H, Hamann T, Offringa R, Jurgens G. Efflux-dependent auxin gradients establish the apical-basal axis of Arabidopsis. Nature 2003;426:147–53.

13. Berleth T, Krogan NT, Scarpella E. Auxin signals – turning genes on and turning cells around. Curr Opin Plant Biol 2004;7:553–63.

14. Jenik PD, Barton MK. Surge and destroy: the role of auxin in plant embryogenesis. Development 2005;132:3577–85.

15. Friml J, Benfey P, Benkova E, Bennett M, Berleth T, Geldner N, Grebe M, Heisler M, Hejatko J, Jurgens G, Laux T, Lindsey K, Lukowitz W, Luschnig C, Offringa R, Scheres B, Swarup R, Torres-Ruiz R, Weijers D, Zazimalova E. Apical-basal polarity: why plant cells don't stand on their heads. Trends Plant Sci 2006;11: 12–14.

16. Friml J, Benkova E, Blilou I, Wisniewska J, Hamann T, Ljung K, Woody S, Sandberg G, Scheres B, Jurgens G. AtPIN4 mediates sink-driven auxin gradients and root patterning in Arabidopsis. Cell 2002;108:661–73.

17. Mayer U, Ruiz RAT, Berleth T, Miseera S, Juurgens G. Mutations affecting body organization in the Arabidopsis embryo. Nature 1991;353:402–7.

18. Mayer U, Buttner G, Jurgens G. Apical-basal pattern formation in the Arabidopsis embryo: studies on the role of the gnom gene. Development 1993;117:149–62.

19. Geldner N, Friml J, Stierhof Y-D, Jurgens G, Palme K. Auxin transport inhibitors block PIN1 cycling and vesicle trafficking. Nature 2001;413:425–8.

20. Geldner N, Anders NWH, Keicher J, Kornberger W, Muller P, Delbarre A, Ueda T, Nakano A, Jurgens G. The Arabidopsis GNOM ARF-GEF mediates endosomal recycling, auxin transport, and auxin-dependent plant growth. Cell 2003;112: 219–30.

21. Lukowitz W, Mayer U, Jurgens G. Cytokinesis in the Arabidopsis embryo Involves the syntaxin-related *KNOLLE* gene product. Cell 1996;84:61–71.
22. Mansfield SG, Briarty LG, Erni S. Early embryogenesis in *Arabidopsis thaliana*. I. The mature embryo sac. Can J Bot 1990;69:447–60.
23. Torres-Ruiz RA, Lohner A, Jurgens G. The *GURKE* gene is required for normal organization of the apical region in the Arabidopsis embryo. Plant J 1996;10:1005–16.
24. Kajiwara T, Furutani M, Hibara K-I, Tasaka M. The *GURKE* gene encoding an acetyl-CoA carboxylase is required for partitioning the embryo apex into three subregions in Arabidopsis. Plant Cell Physiol 2004;45:1122–8.
25. Long JA, Woody S, Poethig S, Meyerowitz EM, Barton MK. Transformation of shoots into roots in Arabidopsis embryos mutant at the *TOPLESS* locus. Development 2002;129:2797–806.
26. Long JA, Ohno C, Smith ZR, Meyerowitz EM. TOPLESS regulates apical embryonic fate in Arabidopsis. Science 2006;312:1520–3.
27. Barton MK, Poethig RS. Formation of the shoot apical meristem in *Arabidopsis thaliana*: an analysis of development in the wild type and in the *shoot meristemless* mutant. Development 1993;119:823–31.
28. Laux T, Mayer KF, Berger J, Jurgens G. The *WUSCHEL* gene is required for shoot and floral meristem integrity in Arabidopsis. Development 1996;122:87–96.
29. Aida M, Ishida T, Fukaki H, Fujisawa H, Tasaka M. Genes involved in organ separation in Arabidopsis: an analysis of the *cup-shaped cotyledon* mutant. Plant Cell 1997;9:841–57.
30. Clark SE, Running MP, Meyerowitz EM. *CLAVATA1*, a regulator of meristem and flower development in Arabidopsis. Development 1993;119:397–418.
31. Clark SE, Running MP, Meyerowitz EM. *CLAVATA3* is a specific regulator of shoot and floral meristem development affecting the same processes as *CLAVATA1*. Development 1995;121:2057–67.
32. Kayes JM, Clark SE. *CLAVATA2*, a regulator of meristem and organ development in Arabidopsis. Development 1998;125:3843–51.
33. Mayer KFX, Schoof H, Haecker A, Lenhard M, Jurgens G, Laux T. Role of WUSCHEL in regulating stem cell fate in the Arabidopsis shoot meristem. Cell 1998;95:805–15.
34. Clark SE. Meristems: start your signaling. Curr Opin Plant Biol 2001;4:28–32.
35. Fletcher JC. Shoot and floral meristem maintenance in Arabidopsis. Annu Rev Plant Biol 2002;53:45–66.
36. Long JA, Moan EI, Medford JI, Barton MK. A member of the KNOTTED class of homeodomain proteins encoded by the *STM* gene of Arabidopsis. Nature 1996;379:66–9.
37. Lenhard M, Jurgens G, Laux T. The *WUSCHEL* and *SHOOTMERISTEMLESS* genes fulfil complementary roles in Arabidopsis shoot meristem regulation. Development 2002;129:3195–206.
38. Byrne ME, Barley R, Curtis M, Arroyo JM, Dunham M, Hudson A, Martienssen RA. *Asymmetric leaves1* mediates leaf patterning and stem cell function in Arabidopsis. Nature 2000;408:967–71.

39. Aida M, Ishida T, Tasaka M. Shoot apical meristem and cotyledon formation during Arabidopsis embryogenesis: interaction among the *CUP-SHAPED COTYLEDON* and *SHOOT MERISTEMLESS* genes. Development 1999;126: 1563–70.

40. Takada S, Hibara K, Ishida T, Tasaka M. The *CUP-SHAPED COTYLEDON1* gene of Arabidopsis regulates shoot apical meristem formation. Development 2001;128:1127–35.

41. Vroemen CW, Mordhorst AP, Albrecht C, Kwaaitaal MACJ, de Vries SC. The *CUP-SHAPED COTYLEDON3* gene is required for boundary and shoot meristem formation in Arabidopsis. Plant Cell 2003;15:1563–77.

42. Treml BS, Winderl S, Radykewicz R, Herz M, Schweizer G, Hutzler P, Glawischnig E, Ruiz RAT. The gene *ENHANCER OF PINOID* controls cotyledon development in the Arabidopsis embryo. Development 2005;132: 4063–74.

43. Schrick K, Mayer U, Horrichs A, Kuhnt C, Bellini C, Dangl J, Schmidt J, Jurgens G. FACKEL is a sterol C-14 reductase required for organized cell division and expansion in Arabidopsis embryogenesis. Genes Dev 2000;14:1471–84.

44. Schrick K, Mayer U, Martin G, Bellini C, Kuhnt C, Schmidt J, Jurgens G. Interactions between sterol biosynthesis genes in embryonic development of Arabidopsis. Plant J 2002;31:61–73.

45. Souter M, Topping J, Pullen M, Friml J, Palme K, Hackett R, Grierson D, Lindsey K. *Hydra* mutants of Arabidopsis are defective in sterol profiles and auxin and ethylene signaling. Plant Cell 2002;14:1017–31.

46. Topping J, May V, Muskett P, Lindsey K. Mutations in the *HYDRA1* gene of Arabidopsis perturb cell shape and disrupt embryonic and seedling morphogenesis. Development 1997;124:4415–24.

47. Diener AC, Li H, Zhou WX, Whoriskey WJ, Nes WD, Fink GR. *STEROL METHYLTRANSFERASE 1* controls the level of cholesterol in plants. Plant Cell 2000;12:853–70.

48. Sakurai A, Fujioka S. Studies on biosynthesis of brassinosteroids. Biosci Biotechnol Biochem 1997;61:757–62.

49. Schrick K, Fujioka S, Takatsuto S, Stierhof YD, Stransky H, Yoshida S, Jurgens G. A link between sterol biosynthesis, the cell wall, and cellulose in Arabidopsis. Plant J 2004;38:227–43.

50. Peng L, Kawagoe Y, Hogan P, Delmer D. Sitosterol-beta-glucoside as primer for cellulose synthesis in plants. Science 2002;295:147–50.

51. Berleth T, Jurgens G. The role of the *MONOPTEROUS* gene in organising the basal body region of the Arabidopsis embryo. Development 1993;118:575–87.

52. Hamann T, Mayer U, Jurgens G. The auxin-insensitive *bodenlos* mutation affects primary root formation and apical-basal patterning in the Arabidopsis embryo. Development 1999;126:1387–95.

53. Hobbie L, McGovern M, Hurwitz LR, Pierro A, Liu NY, Bandyopadhyay A, Estelle M. The *axr6* mutants of *Arabidopsis thaliana* define a gene involved in auxin response and early development. Development 2000;127:23–32.

54. Hardtke CS, Berleth T. The Arabidopsis gene *MONOPTEROS* encodes a transcription factor mediating embryo axis formation and vascular development. EMBO J 1998;17:1405–11.

55. Hamann T, Benkova E, Baurle I, Kientz M, Jurgens G. The Arabidopsis *BODENLOS* gene encodes an auxin response protein inhibiting MONOPTEROS-mediated embryo patterning. Genes Dev 2002;16:1610–5.

56. Hardtke CS, Ckurshumova W, Vidaurre DP, Singh SA, Stamatiou G, Tiwari SB, Hagen G, Guilfoyle TJ, Berleth T. Overlapping and non-redundant functions of the Arabidopsis auxin response factors MONOPTEROS and NONPHOTOTROPIC HYPOCOTYL 4. Development 2004;131:1089–100.

57. Weijers D, Benkova E, Jäger KE, Schlereth A, Hamann T, Kientz M, Wilmoth JC, Reed JW, Jürgens G. Developmental specificity of auxin response by pairs of ARF and Aux/IAA transcriptional regulators. EMBO J 2005;24:1874–85.

58. Hellmann H, Hobbie L, Chapman A, Dharmasiri S, Dharmasiri N, Pozo CD, Reinhardt D, Estelle M. Arabidopsis *AXR6* encodes CUL1 implicating SCF E3 ligases in auxin regulation of embryogenesis. EMBO J 2003;22:3314–25.

59. Aida M, Beis D, Heidstra R, Willemsen V, Blilou I, Galinha C, Nussaume L, Noh Y-S, Amasino R, Scheres B. The *PLETHORA* genes mediate patterning of the Arabidopsis root stem cell niche. Cell 2004;119:109–20.

60. Sabatini S, Heidstra R, Wildwater M, Scheres B. SCARECROW is involved in positioning the stem cell niche in the Arabidopsis root meristem. Genes Dev 2003;17:354–8.

61. Scheres B, Di Laurenzio L, Willemsen V, Hauser MT, Janmaat K, Weisbeek P, Benfey PN. Mutations affecting the radial organisation of the Arabidopsis root display specific defects throughout the embryonic axis. Development 1995;121:53–62.

62. Willemsen V, Wolkenfelt H, de Vrieze G, Weisbeek P, Scheres B. The *HOBBIT* gene is required for formation of the root meristem in the Arabidopsis embryo. Development 1998;125:521–31.

63. Blilou I, Frugier F, Folmer S, Serralbo O, Willemsen V, Wolkenfelt H, Eloy NB, Ferreira PCG, Weisbeek P, Scheres B. The Arabidopsis *HOBBIT* gene encodes a CDC27 homolog that links the plant cell cycle to progression of cell differentiation. Genes Dev 2002;16:2566–75.

64. Abe M, Takahashi T, Komeda Y. Identification of a cis-regulatory element for L1 layer-specific gene expression, which is targeted by an L1-specific homeodomain protein. Plant J 2002;26:487–94.

65. Abe M, Katsumata H, Komeda Y, Takahashi T. Regulation of shoot epidermal cell differentiation by a pair of homeodomain proteins in Arabidopsis. Development 2003;130:635–43.

66. Lauber MH, Waizenegger I, Steinmann T, Schwarz H, Mayer U, Hwang I, Lukowitz W, Jurgens G. The Arabidopsis KNOLLE protein is a cytokinesis-specific Syntaxin. J Cell Biol 1997;139:1485–93.

67. Waizenegger I, Lukowitz W, Assaad F, Schwarz H, Jurgens G, Mayer U. The Arabidopsis *KNOLLE* and *KEULE* genes interact to promote vesicle fusion during cytokinesis. Curr Biol 2000;10:1371–4.

68. Vroemen CW, Langeveld S, Mayer U, Ripper G, Jurgens G, Van Kammen A, de Vries SC. Pattern formation in the Arabidopsis embryo revealed by position-specific lipid transfer protein gene expression. Plant Cell 1996;8:783–91.

69. Mahonen AP, Bonke M, Kauppinen L, Riikonen M, Benfey PN, Helariutta Y. A novel two-component hybrid molecule regulates vascular morphogenesis of the Arabidopsis root. Genes Dev 2000;14:2938–43.

70. Inoue T, Higuchi M, Hashimoto Y, Seki M, Kobayashi M, Kato T, Tabata S, Shinozaki K, Kakimoto T. Identification of CRE1 as a cytokinin receptor from Arabidopsis. Nature 2001;409:1060–3.

71. Di Laurenzio L, Wysocka-Diller J, Malamy J, Pysh L, Helariutta Y, Freshour G, Hahn M, Feldmann K, Benfey P. The *SCARECROW* gene regulates an asymmetric cell division that is essential for generating the radial organization of the Arabidopsis root. Cell 1996;86:423–33.

72. Helariutta Y, Fukaki H, Wysocka-Diller J, Nakajima K, Jung J, Sena G, Hauser M, Benfey P. The *SHORTROOT* gene controls radial patterning of the Arabidopsis root through radial signaling. Cell 2000;101:555–67.

73. Wysocka-Diller JW, Helariutta Y, Fukaki H, Malamy JE, Benfey PN. Molecular analysis of SCARECROW function reveals a radial patterning mechanism common to root and shoot. Development 2000;127:595–603.

74. Nakajima K, Sena G, Nawy T, Benfey PN. Intercellular movement of the putative transcription factor SHR in root patterning. Nature 2001;413:307–11.

2

Maize Embryogenesis

Pilar Fontanet and Carlos M. Vicient

Summary

Plant embryo development is a complex process that includes several coordinated events. Maize mature embryos consist of a well-differentiated embryonic axis surrounded by a single massive cotyledon called scutellum. Mature embryo axis also includes lateral roots and several developed leaves. In contrast to *Arabidopsis*, in which the orientation of cell divisions are perfectly established, only the first planes of cell division are predictable in maize embryos. These distinctive characteristics joined to the availability of a large collection of embryo mutants, well-developed molecular biology and tissue culture tools, an established genetics and its economical importance make maize a good model plant for grass embryogenesis. Here, we describe basic concepts and techniques necessary for studying maize embryo development: how to grow maize in greenhouses and basic techniques for in vitro embryo culture, somatic embryogenesis and in situ hibridization.

Key Words: Greenhouse; embryo rescue; somatic embryogenesis; in situ hybridization.

1. Introduction

Maize (*Zea mays* L.) is a well-suited species for the analysis of plant embryogenesis. The long genetic tradition of maize has unravelled many facets of biochemistry, physiology, gene regulation and developmental control, and extensive collections of mutated lines affected in embryo development are available (*1*). Developed maize embryos can reach more than 70 mg, very large compared with most plant embryos and are thus particularly amenable to study.

From: *Methods in Molecular Biology, vol. 427: Plant Embryogenesis*
Edited by: M. F. Suárez and P. V. Bozhkov © Humana Press, Totowa, NJ

Maize mature embryo consists of an embryo axis and an extensive cotyledon, the scutellum. The embryo axis consists of a plumule, composed by several leave primordia surrounded by the coleoptile, and a primary root surrounded by the coleorhiza. The whole maize embryo development takes about 60 days.

2. Materials

2.1. Growing Maize in a Greenhouse

1. Growth substrate: mix of organic soil and coconut fiber giving a low density, high air flow and pH 5.5–6.0, containing N 250 mg/L, P_2O_5 150 mg/L and K_2O 270 mg/L.
2. Biodegradable pots for germination: 6.5 cm diameter, 9 cm depth. Approximately 4–5 pots can be filled with 1 L of substrate.
3. Nutrient solution, Hoagland's medium (2) modified by Johnson and col. (3): KNO_3 0.85 g/L, NH_4NO_3 0.096 g/L, K_2HPO_4 0.209 g/L, KH_2PO_4 0.489 g/L, $Ca(NO_3)_2 \cdot 4H2O$ 0.596 g/L, $MgSO_4 \cdot 7H_2O$ 0.172 g/L and $FeSO_4 \cdot 7H_2O$ 0.166 g/L.
4. Greenhouse conditions: 28 °C day/20 °C night, 60% relative humidity, >10,000 lux light intensity, 14-h photoperiod. Optimal conditions depend on the lines used.
5. Pots for growing: 28 cm diameter, 23 cm depth. Approximately each pot need to be filled 10 L of substrate.

2.2. Immature Embryo Culture

1. Rescue culture media: Murashige and Skoog Basal Salt Mixture with sucrose 20 g/L, myoinositol 100 mg/L, nicotinic acid 0.4 mg/L, thiamine HCl 0.2 mg/L and agar 8 g/L, pH 5.8 adjusted with KOH.
2. Growth chamber: 26 °C day/ 20 °C night, 60% relative humidity, >10,000 lux light intensity, 16-h photoperiod.

2.3. Somatic Embryogenesis

1. Initiation medium: Murashige and Skoog Basal Salt Mixture, Murashige and Skoog vitamins, sucrose 3%, 2.4-D 5 mg/L, casein hydrolysate 100 mg/L, agar 8 g/L or gelrite 2 g/L. Adjust pH to 5.8.
2. Growth chamber: 26 °C day/ 20 °C night, 60% relative humidity, >10,000 lux light intensity with 16-h photoperiod (use light only during somatic embryo germination).
3. Induction medium: Murashige and Skoog Basal Salt Mixture, Murashige and Skoog vitamins, sucrose 2%, proline 25 mM, casein hydrolysate 100 mg/L, agar 8 g/L or gelrite 2 g/L. Adjust pH to 5.8.

4. Regeneration medium: Murashige and Skoog Basal Salt Mixture, Murashige and Skoog vitamins, sucrose 6%, NAA 1 mg/L, agar 8 g/L or gelrite 2 g/L. Adjust pH to 5.8.

2.4. In Situ Hybridization

1. Restriction enzyme.
2. Tris-EDTA (TE) buffer: 10 mM Tris–HCl pH 7.5, 1 mM ethylene diamine tetraacetic acid (EDTA) pH 8.0.
3. Spectrophotometer.
4. DIG RNA labelling kit (Roche).
5. RNAse inhibitor.
6. RNAse-free DNAse I.
7. Yeast tRNA.
8. Carbonate buffer: mix 30 μL of 100 mM Na_2CO_3 and 20 μL 100 mM $NaHCO_3$ (pH 10.2).
9. Positively charged nylon membrane.
10. Tris-buffered saline (TBS) buffer: 400 mM NaCl, 100 mM Tris–HCl pH 7.5.
11. Blocking reagent (Roche).
12. Anti-DIG antibody conjugated to alkaline phosphatase.
13. Detection buffer-1: 100 mM NaCl, 100 mM Tris–HCl pH 9.5.
14. DIG labelled test strip (Roche).
15. NBT/BCIP stock solution (Roche).
16. Fixative solution: ethanol–formaldehyde–acetic acid in an 80:3.5:5.5 proportion (v/v/v).
17. Histo-Clear II (National Diagnostics).
18. Paraplast.
19. Rotatory microtome.
20. ChemMate™ Capillary Gap Microscope Coated Slides (Dako Cytomation).
21. Slide heater laboratory equipment.
22. Chromium trioxide, 40% (w/v).
23. Poly-D-lysine solution: 50 μg/mL poly-D-lysine in 10 mM Tris–HCl pH 8.0.
24. Oven, 80 °C.
25. Sodium chloride sodium citric acid buffer (SSC): 150 mM NaCl, 15 mM trisodium citrate.
26. Proteinase K.
27. Proteinase buffer, 50 mM EDTA, 100 mM Tris–HCl pH 8.0.
28. Phosphate-buffered saline (PBS) buffer: 137 mM NaCl, 2.7 mM KCl, 4.3 mM Na_2HPO_4, 1.8 mM KH_2PO_4 pH 7.2.
29. Oven, 50 °C.
30. Hybridization buffer: 6× sodium chloride sodium citric acid buffer, 3% sodium dodecyl sulfate, 50% formamide, 100 μg/mL tRNA.
31. Washing buffer: 2× sodium chloride sodium citric acid buffer, 50% formamide.
32. NTE buffer: 0.5 M NaCl, 10 mM Tris–HCl pH 7.5, 1 mM EDTA.
33. RNAse A.
34. Detection buffer-2: 50 mM $MgCl_2$, 100 mM NaCl, 100 mM Tris–HCl pH 9.5.

3. Methods

3.1. Growing Maize in a Greenhouse

Maize grows best and is more easily handled in the field but, if is necessary, maize can be also grown in pots.

1. Distribute growth substrate in germination pots and saturate with nutrient solution until the surface of the substrate is wet and allow it to drenage.
2. Put one seed per pot about 1 cm depth.
3. Put the pots in the greenhouse.
4. Water pots every 2–3 days with nutrient solution until saturation.
5. Wait until four leaves are fully developed. Under adequate lighting, temperature, watering and nutrition, seeds of most of the maize lines emerge from substrate in no more than 5 days after planting, and leaves are developed after about 15 days.
6. Prepare the growing pots watered until the surface of the substrate is damp and allow it to drenage. Put one plantlet per pot.
7. Water pots every day with nutrient solution (*see* **Notes 1** and **2**).

If you suspect that seeds are not in good conservation conditions, it is advisable to sterilize them using the following procedure (*see* **Note 3**):

1. Submerge 5 min in 100% ethanol under shaking.
2. Submerge 7 min in 2% calcium hypochlorite under shaking.
3. Wash four times in distillate sterile water.

3.2. Immature Embryo Culture

Embryo culture allows performing chemical or physiological treatments with developing embryos and is also useful for embryo rescue. Embryo rescue is the in vitro germination of immature embryos that for some reason cannot germinate under normal conditions *(4)*. Homozygous embryo mutants are often lethal, but sometimes lethality is delayed until the last period of embryo development. In these cases, embryo rescue allows recovery of the homozygous mutant lines.

1. Dissect immature ear from plants (*see* **Note 4**).
2. Remove bracteas and pistils.
3. Clean the ear with water and immerse 2 min in 70% ethanol. Rinse in sterile water.
4. Carefully remove the kernels under sterile conditions by cutting the pedicel using a surgical blade.
5. Sterilize the kernels with 100% ethanol for 5 min and then 10 min in 20% sodium hypochlorite. Rinse at least three times with sterile water.
6. Extract the embryo by placing the kernels on a sterile dish, embryo side facing up, holding the kernel gently with sterile forceps, cutting the pedicel as close to the embryo as possible avoiding the embryo tip and pressing gently the kernel

from top to basis in the embryo side using a sterile surgical blade tip. The embryo will slip out of the kernel.

7. Place the embryo immediately on the rescue culture media plate with the embryo axis up (scutellum side in contact with the media).

8. Keep embryo on dark at the growth chamber until germination (4–5 days). Then transfer germinated embryos to a 16-h light/8-h dark photoperiod.

9. At two to three leaf stage, and provided the root system has reasonably developed, transplant seedling into appropriate soil mixture (*see* **Subheading 3.1**). Keep them covered with plastic wrap during the first days after transplanting. Gradually make holes on wrap over the following days, and finally remove the wrap when seedling is acclimated and begin to grow vigorously.

3.3. Somatic Embryogenesis

Somatic embryogenesis represents the production of embryo-like structures from somatic cells, without gamet fusion. The development of the somatic embryos recapitulates several of the morphologic and developmental events that occur during zygotic embryogenesis making them an interesting experimental model for studying the early stages of plant embryo development *(5,6)*. Although somatic embryos arise naturally in some species, usually somatic embryogenesis requires two steps. First, a high concentration of auxin stimulates cell dedifferentiation. Second, reduction of auxin concentration in the culture media induces differentiation of the undifferentiated cells into somatic embryos. Following an appropriate treatment, somatic embryos can generate a complete plant.

Several factors are critical for the production of maize somatic embryos, including explant source, physiological status, developmental stage and genotype. In maize, immature embryos are the most suitable explant for the establishment of embryogenic cultures *(7)*. Best response is obtained from embryos in which morphological development has been nearly completed. In maize, the genotype strongly influences somatic embryogenesis *(8)*. The following protocol have been optimized for A188 immature maize embryos harvested 14 days after pollination. Conditions for other lines and/or stages must be determined empirically.

1. Sterile immature embryos obtained as indicated for the embryo rescue protocol (*see* **Subheading 3.2**) are placed on initiation medium in Petri plates. It is very important to place the embryos with the embryo axis in contact with the medium. Place 16–20 embryos in each 10-cm Petri plate containing 20–25 mL of medium (or six to seven embryos in a 6-cm Petri plate with 10 mL of medium).

2. Incubate the plates at 26 °C in the dark. Callus become visible after about 7 days.

3. After 4 weeks, separate the type II embryogenic callus (*see* **Note 5**). A subculture routine of 2–4 weeks is recommended, but care should be taken at the time of

each subculture to exclude non-embryogenic segments. Longer subculture periods may reduce regeneration ability because of the formation of increased amounts of non-embryogenic callus.

4. In order to induce the production of somatic embryos, transfer a portion of embryogenic callus to plates containing induction medium and incubate at 26 °C in the dark. Structures resembling coleoptiles will appear about 10–15 days later.

5. In order to regenerate adult plants, germinating somatic embryos should be placed on regeneration medium. Culture in the light at 28 °C until plants are well rooted and shoots have elongated to the top of the box (1–2 weeks).

6. Transfer plants to soil after the formation of a well-developed shoot and root system, and grow them to maturity, initially in a humidified chamber for about a week before transfer to the greenhouse (*see* **Subheading 3.2**).

3.4. In Situ Hybridization

In situ hybridization techniques enable the analysis of RNA accumulation in individual cells and have been extremely useful in development studies *(9)*.

3.4.1. Probe Labeling

Single-stranded RNA probes are usually used for in situ hybridizations since they allow the removal of unbound probe after hybridization without signal loss. The most commonly used label for in situ hybridizations is DIG-11-UTP.

1. Linearlize 5 μg plasmid DNA. Linearize the probe as follows: sense probe using a restriction enzyme which cuts 3′ of the sequence you want to use, antisense probe using a restriction enzyme which cuts 5′ of the sequence. Check for complete digestion on agarose gel.

2. Add 2.5–3 volumes of 95% ethanol/0.12 M sodium acetate, invert to mix and incubate in an ice-water bath for at least 10 min.

3. Centrifuge at 12,000 g for 15 min at 4 °C and decant the supernatant.

4. Add 80% ethanol (corresponding to about two volumes of the original sample), incubate at room temperature for 5 min and centrifuge again for 5 min, remove the supernatant and air-dry the pellet.

5. Dissolve dried DNA in no more than 13 μL TE. Check concentration in spectrophotometer (*see* **Note 6**).

6. Mix the following components of DIG RNA labeling kit in this order:
   a. A volume of linearized DNA corresponding to 2 μg (× μL).
   b. (13 - ×) μL water.
   c. 2 μL 10× NTP labelling mixture.
   d. 2 μL transcription buffer (10×).
   e. 1 μL RNAse inhibitor.
   f. 2 μL (40 U) DNA-dependent RNA polymerase (T7/SP6/T3 depending on both the orientation of the cDNA and the vector used) (*see* **Note 7**).

7. Incubate at 37 °C for 2 h.
8. Add 2 µL (20 U) of RNAse-free DNAse I and incubate at 37 °C for 15 min.
9. Add 2 µL of 0.2 M EDTA (pH 8.0) (*see* **Note 8**).
10. Analyze 1 µL of the reaction in agarose gel (*see* **Note 9**).
11. Add: 1 µL of 10 µg/µL yeast tRNA, 37.5 µL of 5 M ammonium acetate, 24 µL of water and 220 µL of cold ethanol (–20 °C).
12. Maintain for 30 min at –80 °C or 1 h at –20 °C.
13. Collect the RNA pellet (13,000 g, 10 min).
14. Wash the pellet with ice-cold 70% ethanol (v/v) and centrifuge (13,000 g, 10 min).
15. Remove the supernatant and air-dry.
16. Dilute in 10 µL water.
17. Remove 1 µL to check the probe by a dot-blot (*see* **Subheading 3.4.2**). Store the rest at –20 °C.

The optimal probe size for in situ hybridization has been estimated between 100 and 250 bp; however, probes as large as 600 bp have been successfully used. If larger probes are to be used they must be partially hydrolysed by alkaline hydrolysis.

1. Substitute **step 7** of labeling protocol by adding 50 µL of carbonate buffer (*see* **Note 10**).
2. Incubate at 60 °C for the appropriate time (*see* **Note 11**).
3. Add the following:

   a. 1 µL of yeast tRNA (1 µg/µL).
   b. 5 µL of 5% (v/v) acetic acid.
   c. 5 µL of 3 M sodium acetate pH 4.5.
   d. 250 µL cold 100% ethanol.

4. Leave at –80 °C for 30 min or 1 h at –20 °C.
5. Collect the RNA pellet (13,000 g, 10 min).
6. Wash the pellet with 70% cold ethanol (–20 °C) and centrifuge (13,000 g, 10 min).
7. Remove the supernatant and air-dry.
8. Dilute in 10 µL water.
9. Keep 1 µL to check DIG incorporation and store at –20 °C.

3.4.2. Dot-Blot Analysis of Probe Labeling

Probe labeling efficiency is estimated using Ab detection with a DIG-labelled test trip.

1. Use 1 µL of probe to prepare the following dilutions: 1/25, 1/250 and 1/2500 (in water).
2. Spot 1 µL of each dilution onto a nylon membrane and UV cross-link it (30 mJ).
3. Wet in TBS buffer. Do the same in parallel with a DIG-labelled test trip.

4. Incubate 30 min at gentle shaking in TBS buffer containing 0.5% (w/v) blocking reagent (*see* **Note 12**).
5. Wash in TBS buffer.
6. Incubate 15 min in 9 mL of TBS buffer containing 1 µL of anti-digoxigenin-alkaline phosphatase conjugate (*see* **Note 13**) agitating gently.
7. Rinse for 15 min in detection buffer 1.
8. Incubate in 10 mL of detection buffer 1 containing 200 µL of NBT/BCIP stock solution.
9. Signal should be detected within a few minutes. Wait until you can see a signal in all dots.
10. Wash membrane in TBS buffer.
11. Dilute the probes to the correct dilution rate with hybridization buffer (*see* **Note 14**).

3.4.3. Fixation, Embedding and Sectioning

Fixation stabilizes the tissue structure prior to subsequent treatment, while retaining the physical and biochemical characteristics of the living material.

1. Cut kernels or dissected embryos in small pieces (not bigger than 3 mm in width) (*see* **Note 15**).
2. Submerge the samples in fresh fixative solution for 1 h at room temperature (*see* **Notes 16** and **17**).
3. Discard and replace by fresh fixative solution and leave for 1 week at 4 °C.
4. Replace the fixative solution by 70% ethanol. The samples can be stored at 4 °C for at least 1 year.
5. Incubate successively at room temperature in the following solutions for at least 2 h each step:

   a. 70% ethanol.
   b. 80% ethanol.
   c. 90% ethanol (*see* **Note 18**).
   d. Twice in 100% ethanol.
   e. 75% ethanol/25% Histo-Clear II (*see* **Note 19**).
   f. 50% ethanol/50% Histo-Clear II.
   g. 25% ethanol/75% Histo-Clear II.
   h. Twice in 100% Histo-Clear II.

6. Add one volume of paraplast wax beans and leave overnight at room temperature.
7. Add 0.25 volumes of melted paraplast and incubate at 42 °C for at least 3 h (*see* **Note 20**).
8. Pour off the Histo-Clear II/paraplast solution and add pure melted paraplast. Incubate for at least 6 h at 42 °C. Repeat at least four times.
9. Pour the molten paraplast and tissue sample into a pre-warm mould at 42 °C. Align section in mould using warmed forceps.

10. Put the mould on cold water (about 18 °C) for about 1 min and allow to cool overnight at 4 °C before sectioning (*see* **Note 21**).
11. Trim away excess paraplast and cut around the section to form a trapezoid with the smallest face presenting the surface for sectioning.
12. Use a standard rotatory microtome for paraplast sections of 8–12 μm thickness using a metal knife. During sectioning, store the next paraplast block on ice.
13. Paraplast sections come off in ribbons. Once cut, sections are first spread on a 45 °C water bath.
14. Mount sections directly on coated slides (*see* **Note 22**).
15. Leave on a heat surface (42 °C) overnight.
16. Store at 4 °C in a dry dust-free box until use.

Commercial coated slides are available but they also can be home-made.

1. Clean slides by incubating in 40% (w/v) chromium trioxide overnight.
2. Wash with distilled water for 10 min at least 10 times.
3. Bake at 80 °C overnight.
4. Coat the slides with poly-D-lysine solution (*see* **Note 23**).
5. Leave slides to air dry in a dust-free place.
6. Store at room temperature in a dry box. Treated slides can be kept at room temperature for 1 year.

3.4.4. Pre-Hybridization Treatments

Pre-hybridization make the target RNA more accessible. This is achieved by a proteinase K digestion. The optimal deproteinization time varies depending on the tissue, fixation and protease batch. We indicate an average time, but it is recommended to calibrate the optimal time for your own conditions and tissues.

1. Incubate the slides 2 × 2 min in Histo-Clear II.
2. Incubate 2 min in the following:

    a. 100% ethanol (twice).
    b. 95% ethanol (v/v in distillate water).
    c. 80% ethanol.
    d. 70% ethanol.
    e. 50% ethanol.
    f. 30% ethanol.
    g. Distillate water (twice).

3. Incubate for 20 min in 0.2 M HCl.
4. Rinse in distillate water 5 min.
5. Transfer to 2× sodium chloride sodium citric acid buffer for 5 min.
6. Rinse again in distillate water 5 min.
7. Incubate the sections for 30 min (*see* **Note 24**) at room temperature in 1 μg/mL proteinase K in proteinase buffer.

8. Rinse in 2 mg/mL glycine in PBS buffer for 2 min.
9. Rinse in PBS for 2 min.
10. Incubate in 4% (v/v) formaldehyde in PBS buffer for 10 min (*see* **Note 25**).
11. Wash in PBS buffer twice for 5 min each.
12. Incubate 2 min each in the following:

    a. Distillate water.
    b. 50% ethanol.
    c. 70% ethanol.
    d. 95% ethanol.
    e. 100% ethanol (twice).

13. Dry in a dessicator for 20 min. (*see* **Note 26**).

3.4.5. Hybridization and Washes

Hybridization is conducted at 50 °C in a buffer containing 50% (v/v) formamide. Formamide permits the washes to be stringent enough without having to increase the temperature above 50 °C which could damage the tissue. To decrease the background, RNAse A digestion is used. RNAse A digests only ssRNA therefore removing the unbound probe without decreasing the signal.

1. Warm slides at 50 °C.
2. Prepare 400 μL diluted probe per pair of slides in hybridization buffer at the appropriate concentration.
3. Denature probes at 80 °C for 2 min. Spin and put them on ice.
4. Apply the probe to sections. Cover with another slide, in pairs.
5. Incubate at 50 °C in a humidified box for at least 17 h (*see* **Note 27**).
6. Wash twice for 90 min each in washing buffer at 50 °C.
7. Incubate in NTE buffer at 37 °C for 5 min.
8. Incubate in NTE buffer containing 20 μg/mL RNAse A at 37 °C for 30 min.
9. Wash five times for 10 min with NTE buffer at 37 °C.
10. Incubate for 1 h in washing buffer at 50 °C.
11. Wash in PBS buffer for 5 min.
12. Replace with fresh PBS buffer.

3.4.6. DIG Detection

1. Wash the slides for 5 min in TBS buffer.
2. Incubate 1 h in 0.5% blocking reagent in TBS buffer under gentle shaking (*see* **Note 12**).
3. Wash in 1% bovine serum albumin, 0.3% Triton X-100 in TBS buffer for 30 min.
4. Incubate 90 min in 1% bovine serum albumin, 0.3% Triton X-100 (v/v) in TBS buffer containing 1:3000 dilution of anti-DIG antibody conjugated to alkaline phosphatase (*see* **Note 13**).

5. Wash in 1% bovine serum albumin, 0.3 % Triton X-100 in TBS buffer for 20 min. Repeat three times.
6. Wash for 5 min in detection buffer 2.
7. For each slide: mix 0.3 mL of detection buffer 2 and 5 μL NBT/BCIP stock solution. Scale up for the total number of slides.
8. Incubate slides in the dark during at least 12 h. Check at 12 h and then after each additional 12-h period.
9. Stop the reaction by a series of washes 0.5–5 min each (*see* **Note 28**).

   a. Distillate water.
   b. 70% ethanol.
   c. 100% ethanol.
   d. 70% ethanol.
   e. Distillate water.

10. Dry under vacuum and mount. Store in a dry dust-free box. Avoid prolonged exposure to light.

4. Notes

1. Drop watering is recommended.
2. Under adequate lighting, temperature, watering and nutrition, plants initiate flowering about 9–12 weeks after planting and produce dry seeds 8 weeks after pollination. Exact timing depends on the genotype.
3. Sterilized seeds can be kept at 4 °C in sterile conditions.
4. The optimal age of the ear depends on the line. In general, the older is the plant, the better is germination, but in the mutant lines, this is sometimes not a case. Embryos younger than 13 dap (days after pollination) usually do not germinate.
5. Two types of callus will form on the scutellar surface: type I and type II. Whereas type I callus is composed by well organized compact cells, type II is friable, soft and colorless. Somatic embryos are preferently formed in the embryogenic type II cultures.
6. Do not phenol/chloroform extract because probe will partition into the organic phase.
7. Choose your polymerase as follows: sense probe use a polymerase with promoter 5′ to the sequence you want transcribed, antisense probe use a polymerase with promoter 3′ to the sequence you want transcribed. Antisense probe will detect transcripts and sense probe is used as a control of non-specific signals.
8. A solution containing 50 μg per mL of double-strand DNA has an absorbancy (optical density) of 1.0 at a wave length of 260 nm.
9. RNA, 3–10 μg, are synthesized.
10. Prepare the buffer fresh each time.
11. Time is calculated according to the following formula. Lo is initial transcript length (Kb). L is the final desired length, usually 0.1 Kb. T, time in minutes. $T = (Lo - L)/(0.11 \times Lo \times L)$.

12. Blocking reagent must be prepared fresh just before use and should be dissolved at 60 °C for 30 min. The solution remains turbid.

13. Centrifuge the anti-DIG-AP for 5 min at 13,000 g in the original vial and pipette from the surface.

14. The probe should be used at a dilution close to the most diluted signal on the test strip.

15. The smaller is the sample the better is fixation. If whole kernels are used, extract the pericarp or make cuts in it in order to allow fixative to enter the tissues. It is not recommended to use samples bigger than 3 mm in width.

16. It is essential that fixatives are made up freshly from fresh chemical stocks; self-polymerization reduces the effectiveness of solutions and degradation products from stored stocks cause tissue damage.

17. During the fixation, desiccation and embedding processes, the tissue : solution ratio should always be lower than 1:10.

18. 90% ethanol step can include Fast Green 0.5% (w/v) in order to stain the samples and facilitate posterior mounting.

19. Histo-Clear II could be substituted by xylene, but xylene is toxic by aspiration.

20. Melted paraplast temperature should never be higher than 62 °C at any time as this will damage the polymer structure of the paraplast.

21. Paraplast blocks can be stored at 4 °C in a dehydrated atmosphere for at least 1 year.

22. For short-term storage (no more than one week), place the ribbons into suitable dust-free box.

23. Poly-D-lysine can be prepared in 10× stock in Tris–HCl, pH 8.0, and stored at −20 °C.

24. Incubation time depends on proteinase batch and tissue samples and must be optimized empirically.

25. Thirty milliliters of 37% formaldehyde in 270 mL PBS buffer.

26. At this stage, the sections can be stored in a desiccation box for 2 h at room temperature.

27. When using non-homologous probes, the hybridization temperature must be reduced: 1% sequence divergent lowers the hybridization temperature by 1 °C.

28. Ethanol changes the color of the signal from brown to blue. It is also capable of washing unspecific precipitates, but it can also weaken the signal. If the background is low and the signal high then 5 min in EtOH is enough, but if the signal is low then wash only 30 s in each dilution.

References

1. Van Lijsebettens M, Van Montagu M. Historical perspectives on plant developmental biology. Int J Dev Biol 2005;49:453–465.

2. Hoagland DR, Arnon DL. The water culture method for growing plants without soil. California Agr Expt Sta Circ 1938;347.

3. Johnson CM, Stout PR, Broyer TC, Carlton AB. Comparative chlorine requirements of different plant species. Plant Soil 1957;8:337–353.
4. Consonni G, Aspesi C, Barbante A, Dolfini S, Giuliani A, Hansen S, Brettschneider R, Pilu R, Gavazzi G. Analysis of four maize mutants arrested in early embryogenesis reveals an irregular pattern of cell division. Sex Plant Reprod 2003;15:281–290.
5. Dodeman VL, Ducreux G, Kreis M. Zygotic embryogenesis versus somatic embryogenesis. J Exp Bot 1997;48:1493–1509.
6. Hecht V, Vielle-Calzada JP, Hartog MV, Schmidt EDL, Boutilier K, Grossniklaus U, de Vries SC. The Arabidopsis somatic embryogenesis receptor kinase 1 gene is expressed in developing ovules and embryos and enhances embryogenic competence in culture. Plant Physiol 2001;127:803–816.
7. Morrish FV, Vasil V, Vasil IK. Developmental morphogenesis and genetic manipulation in tissue and cell cultures of Gramineae. Adv Genet 1987;24:431–499.
8. Hodges TK, Kamo KK, Imbrie CW, Becwar MR. Genotype specificity of somatic embryogenesis and regeneration in maize. Bio/Technology 1986;4:219–223.
9. Angerer LM, Angerer RC. Localization of mRNAs by *in situ* hybridization. Methods Cell Biol 1991;35:37–71.

3

Spruce Embryogenesis

Sara von Arnold and David Clapham

Summary

Somatic embryogenesis, the process in which embryos, similar in morphology to their zygotic counterparts, are induced to develop in culture from somatic cells, is a suitable model system for investigating the regulation of embryo development. Through this process, a large number of embryos at defined stages of development can easily be obtained. Somatic embryogenesis in Norway spruce is comprised of a sequence of steps including initiation, proliferation, early embryo formation, embryo maturation, desiccation and germination. To execute this pathway, a number of critical physical and chemical treatments should be applied with proper timing. Embryogenic cell lines of Norway spruce are initiated from zygotic embryos. The cell lines proliferate as proembryogenic masses (PEMs) in the presence of auxin and cytokinin. Early somatic embryos develop from PEMs after withdrawal of auxin and cytokinin. PEM to somatic embryo transition is a key developmental switch that determines the yield and quality of mature somatic embryos. The embryos develop further, to a stage corresponding to late embryogeny, in the presence of abscisic acid. Some cell lines deviate from normal pattern formation exhibiting developmental arrest at certain stages. These arrested cell lines, together with transgenic lines, are valuable tools for studying embryo development. Particle bombardment is routinely used to produce transgenic plants of Norway spruce.

Key Words: Conifers; cryopreservation; embryogenesis; gene transfer; gymnosperms; Norway spruce.

1. Introduction

1.1. Zygotic Embryogenesis

Mutagenesis and the subsequent screening for and analysis of mutants, mainly in *Arabidopsis*, has been used for elucidating the genetic regulation of embryonic pattern formation in angiosperms. Such approaches are not applicable

From: *Methods in Molecular Biology, vol. 427: Plant Embryogenesis*
Edited by: M. F. Suárez and P. V. Bozhkov © Humana Press, Totowa, NJ

to gymnosperms, which have large genomes, about 200–400 times bigger than in *Arabidopsis*, large size and a long regeneration time. Molecular data suggest that extant seed plants (gymnosperms and angiosperms) share a last common ancestor about 285 million years ago. From an evolutionary point of view, it is important to learn more about the regulation of embryogenesis in gymnosperms. Another reason to study gymnosperms, and especially conifers, is that they are of great commercial importance.

The sequence of embryo development in gymnosperms can be divided into three phases *(1)*. (i) Proembryogeny – all stages before elongation of the suspensor. (ii) Early embryogeny – all stages after elongation of the suspensor and before establishment of the root meristem. (iii) Late embryogeny – establishment of the root and shoot meristems and further development of the embryo following this event. There are four types of proembryogeny in gymnosperms, of which the conifer type is the most common and is interpreted as the basal plan for gymnosperm proembryogeny. Proembryo development begins when the fertilized egg nucleus divides into four free nuclei. After division, two tiers are formed, the primary embryonal tier and the primary upper tier, which become cellular. Internal divisions of these two tiers produce four tiers. The lower two tiers constitute the embryonal tier. The upper four cells of the embryonal tiers elongate to form a functional suspensor and the lower four cells form the embryonal mass. Early embryogeny begins with the elongation of the embryonal suspensor. The cells in the lower embryonal tier divide, creating an embryonal mass. The lack of restriction in cell divisions means that all cells, including the surface layer, continue to divide periclinally as well as anticlinally, preventing differentiation of a distinct protoderm. The basal cells of the embryonal mass continue to divide predominantly in a transverse plane and elongate, contributing to the thick embryonal suspensor. The suspensor cells closest to the embryonal mass have been named embryonal tube cells. Late embryogenesis is a period of histogenesis and organogenesis. Early during this stage, the root and shoot apical meristems are delineated and the plant axis established. The root apical meristem forms near the center of the embryo, first as a root organizing center. The shoot meristem originates at the distal part of the embryonal mass and is relatively superficial compared to the root organizing center. The cotyledon primordia arise in a ring around the distal end of the embryo. At this stage, provascular and cortex tissue differentiate.

Embryo development is difficult to study in conifers owing to the inaccessability of the embryo inside the ovule. Our possibilities to study the mechanisms underlying conifer embryogenesis have been enhanced considerably during the last years by using somatic embryos of Norway spruce as a model.

1.2. Somatic Embryogenesis

The somatic embryo system includes a stereotyped sequence of developmental stages, resembling zygotic embryogeny, that can be synchronized by specific treatments, making it possible to collect a large number of somatic embryos at specific developmental stages. Embryogenic cell lines of Norway spruce are established from zygotic embryos, so that each cell line represents one genotype. The cell lines proliferate as proembryogenic masses (PEMs) in the presence of auxin and cytokinin. Somatic embryos differentiate from PEMs after withdrawal of auxin and cytokinin. This stage corresponds to early embryogeny. The embryos develop further, to a stage corresponding to late embryogeny, in the presence of abscisic acid (ABA). A time-lapse tracking technique has been used to analyse the developmental pathway of somatic embryogenesis in Norway spruce *(2)*. A representative example of how standard embryogenic cultures proliferate and develop is shown in **Fig. 1**. Based on the knowledge gained from studying the developmental pathway of somatic embryogenesis, we have made a model of the process (*see* **Fig. 2**).

In addition to cell lines, that develop somatic embryos as described in **Fig. 1**, there are lines that are arrested at PEM I, PEM II or PEM III. Cell lines arrested at PEM III can sporadically differentiate embryos, but these will not develop normal cotyledonary embryos. Cell lines arrested at PEM I or PEM II do not differentiate somatic embryos, regardless of treatment. Most of the cell lines arrested at PEM I and PEM II continue to proliferate as PEMs also after withdrawal of auxin and cytokinin. Embryo-specific mutants are not available in conifers. Therefore, these arrested cell lines together with transgenic cell lines are valuable tools for studying the regulation of embryo development.

Global changes in gene expression during successive stages of embryo development can be followed by microarray analysis of synchronized embryogenic cultures. RNA extracted from embryogenic cultures of Norway spruce representing successive developmental stages was analysed on DNA microarrays *(3,4)*. A group of 107 genes (29% of arrayed cDNAs) were upregulated upon PEM to embryo transition, then downregulated during early embryogeny and finally upregulated again at the beginning of late embryogeny. The finding of a transcriptionally repressive state during early embryogeny supports the idea that the major principles of genetic regulation of embryogenesis can be conserved between higher plants and metazoan animals.The variation in expression of specific genes suggests that a high level of DNA methylation is important for the differentiation of somatic embryos from PEMs. Furthermore, downregulation of genes involved in auxin metabolism is important for apical–basal polarization during early embryogeny, and downregulation of genes involved in cell wall formation is important for differentiation of primary meristems. These

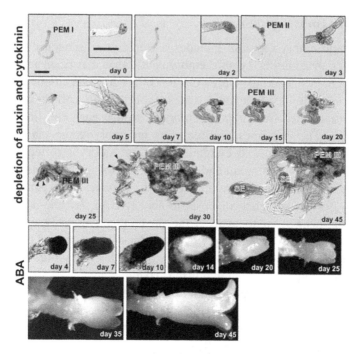

Fig. 1. Time-lapse tracking of somatic embryo formation and development in Norway spruce. The tracking was performed in a thin layer of agarose, first under gradual depletion of auxin and cytokinin and then followed by addition of abscisic acid (ABA) (for details *see* **ref. 2**). The development starts with a small cell aggregate proembryogenic mass (PEM) I composed of a small compact clump of densely cytoplasmatic cells adjacent to a single vacuolated cell showing a tendency to elongation. After 3 days, additional elongated cells have developed from the compact clump of densely cytoplasmic cells forming PEM II. Both PEM I and PEM II are polar structures. However, the polarity is successively lost owing to an increased proliferative activity of densely cytoplasmatic cells. After about 2 weeks, large cell aggregates classified as PEM III are formed. Depletion or withdrawal of auxin and cytokinin stimulates differentation of somatic embryos (SE) from PEM III (day 25 and day 30, arrowhead denotes SE). The first visible response of early SEs to ABA is that the embryonal mass becomes opaque. This response is accompanied by degradation of the suspensor by programmed cell death. Thereafter, the embryo starts to elongate and differentiate cotyledons. (Originally published by **ref. 2**, with the permission Kluwer Academic Publishers).

first microarray data demonstrate how the field of conifer embryology is being expanded by the avaiability of this technology combined with somatic embryos.

Formation of the apical–basal pattern during early embryogeny proceeds through the establishment of three major cell types: the meristematic cells

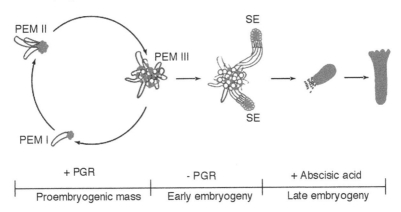

Fig. 2. Schematic representation of the developmental pathway of somatic embryo-genesis in Norway spruce. The process involves two broad phases. The first phase, which occurs in the presence of auxin and cytokinin (+PGR), is represented by prolifer-ating proembryogenic masses (PEMs), cell aggregates which can pass through a series of three characteristic stages distinguished by cellular organization and cell number (PEM I, II and III), but cannot develop into a normal embryo. The second phase encom-passes development of somatic embryos from PEM III and is triggered by withdrawal of auxin and cytokinin (–PGR). Once early somatic embryos have formed, their further development requires abscisic acid. (Originally published by Filonova et al., 2000, reproduced with the permission of Kluwer Academic Publishers).

of the embryonal mass, the embryonal tube cells and terminally differen-tiated suspensor cells. The embryonal mass cells are small and spherical with dense cytoplasm and high mitotic activity. The asymmetric divisions of the most basally situated cells within the embryonal mass give rise to a layer of elongated embryonal tube cells that differentiate to form one layer of the suspensor cells. By this means, reiterated asymmetric divisions in the embryonal mass are continuously adding new layers of cells to the growing suspensor which becomes composed of several layers of highly vacuolated elongated cells *(5)*. The embryo suspensor is a terminally differentiated structure which is eliminated by programmed cell death (PCD) during late embryogeny *(5–7)*. Expression and activation of metacaspase is essential for maintaining the proper balance between cell proliferation and PCD *(7,8)*. The organization of micro-tubules and F-actin changes successively from the embryonal mass towards the distal end of the embryo suspensor *(9)*. The microtubule arrays appear normal in the embryonal mass cells, but are partially disorganized in the embryonal tube cells and disrupted in the suspensor cells. The organization of F-actin gradually changes from a fine network in the embryonal mass cells to thick cables in the suspensor cells. F-actin depolymerization drugs abolish normal embryonic pattern formation and associated PCD in the suspensor. Both microtubules and F-actin are important for embryogenesis, but it is F-actin that is critical for the execution of embryonic PCD.

Differentiation of the outer cell layer in the embryonal mass during early embryogeny is regulated by *PaHB1* (*Picea abies* Homeobox1), whose encoded protein is highly similar to those from the HD-GL2 angiosperm counter-parts *(10)*, suggesting similarities in the definition of the outer cell layer in seed plants. As in angiosperms, proper functioning of the outer cell layer in Norway spruce requires a specific expression pattern of a lipid transfer protein like gene, *Pa18 (11)*. The expression of *PaHB1* and *Pa18* switches from ubiquitous expression in PEMs to an outer cell layer-specific localization in early somatic embryos. Furthermore, ectopic expression of *PaHB1* or *Pa18* leads to an early developmental block. Three members of the *KNOX* (*KNOTTED1*-like homeobox) family (*HBK1*, *HBK2* and *HBK3*) are expressed at all stages during somatic embryo development in Norway spruce *(12)*. However, *HBK2* is not expressed in blocked cell lines, suggesting that expression of *HBK2* is necessary for normal embryo development. The *PtNIP1;1* gene encodes an aquaglycero-porin that is expressed early in embryogenesis. After fusing the promoter of *PtNIP1;1* to the *uidA* gene and transforming it into embryogenic cultures of Norway spruce, transformants exhibited GUS expression that was uniform in PEMs, confined to the suspensor region in early somatic embryos and absent at late embryogeny and in cotyledonary embryos *(13)*.

During late embryogeny, *PaHB2* participates in the maintenance of the radial pattern by specifying cell identity in the cortical layers *(14)*. A Norway spruce homologue of the maize *Viviparous 1* gene (*Pavp1*) is expressed during maturation of somatic embryos *(15)*. Once the early cotyledonary stage is reached, the expression declines. This pattern of *Pavp1* expression during maturation of Norway spruce somatic embryos is similar to that of angiosperm VP1 homologues.

Taken together, the regulation of embryo formation has many similarities in gymnosperms and angiosperms, but there are also differences which have to be studied in more detail.

2. Materials

2.1. The System of Somatic Embryogenesis in Norway Spruce

2.1.1. Initiation

1. Immature cones of Norway spruce collected 2 months after fertilization or mature seeds.
2. Solidified LP medium: 18.8 mM KNO_3, 7.5 mM NH_4NO_3, 1.5 mM $MgSO_4$, 2.5 mM KH_2PO_4, 3.0 mM $CaCl_2$, 50 μM Fe-EDTA, 10 μM Zn-EDTA, 10 μM

$MnSO_4$, 10 μM H_3BO_3, 0,1 μM Na_2MoO_4, 0.01 μM $CuSO_4$, 0.01 μM $CoCl_2$, 4.5 μM KI, 4.9 μM pyridoxine-HCl, 16.3 μM nicotinic acid, 26.7 μM glycine, 15 μM thiamine-HCl, 500 μM *myo*-inositol, 1 mM D-glucose, 1 mM D-xylose, 1 mM L-arabinose, 0.56 μM L-alanine, 0.13 μM L-cysteine-HCl, 0.06 μM L-arginine, 0.08 μM L-leucine, 0.06 μM L-phenylalanine, and0.06 μM L-tyrosine. The medium is diluted to half-strength. The diluted medium is supplemented with 30 mM D-sucrose, 2,4-dichlorophenoxyacetic acid (2,4-D, 9.0 μM), N^6-benzyladenine (BA, 4.4 μM) and solidified with 0.35% (w/v) Gelrite. The pH is adjusted to 5.8 ± 0.1 prior to autoclaving. L-glutamine (3 mM) is filter-sterilized and added into autoclaved and cooled medium. The medium is poured into sterile Petri dishes. The original LP medium was described by von Arnold and Eriksson *(16)*. The adjusted medium was described by Bozhkov and von Arnold *(17)*.

2.1.2. Proliferation

1. Half-strength solidified LP medium (*see* **Subheading 2.1.1**).
2. Half-strength liquid LP medium (the same composition as described in **Subheading 2.1.1.**, except that the concentration of NH_4NO_3 in full strength LP medium before dilution is 15 mM and Gelrite is excluded). Sterile medium is poured into 250-mL Erlenmeyer flasks (97 mL per flask).

2.1.3. Pre-Maturation

1. Half-strength liquid LP medium (the same composition as described in **Subheading 2.1.1.**, except that Gelrite, 2,4-D and BA are omitted).

2.1.4. Maturation

1. Solidified BMI-SI medium: 23 mM KNO_3, 3.5 mM NH_4NO_3, 1.5 mM $MgSO_4$, 0.6 mM KH_2PO_4, 1.5 mM $CaCl_2$, 100 μM Fe-EDTA, 10 μM Zn-EDTA, 5 μM $MnSO_4$, 10 μM H_3BO_3, 0.1 μM Na_2MoO_4, 0.01 μM $CuSO_4$, 0.01 μM $CoCl_2$, 4.5 μM KI, 2.5 μM pyridoxine-HCl, 4 μM nicotinic acid, 26.7 μM glycine, 3 μM thiamine-HCl, 5.5 mM *myo*-inositol, 90 mM D-sucrose and 500 mg/L casein hydrolysate, solidified with 0.35% (w/v) Gelrite. The pH is adjusted to 5.8 ± 0.1 prior to autoclaving. L-glutamine (3 mM) and ABA (30 μM) are filter-sterilized and added into autoclaved and cooled medium. The medium is poured into sterile Petri dishes. The original BMI-SI medium was described by Krogstrup *(18)*.

2.1.5. Germination

1. Solidified SH medium: 2.6 mM $NH_4H_2PO_4$, 1.4 mM $CaCl_2$, 1.5 mM $MgSO_4$, 25 mM KNO_3, 50 μM Fe-EDTA, 79 μM H_3BO_3, 0.4 μM $CoCl_2$, 0.4 μM $CuSO_4$, 20 μM $MnSO_4$, 0.4 μM Na_2MoO_4, 5.6 μM KI, 20 μM Zn-EDTA, 1 mM *myo*-inositol and 60 mM D-sucrose. The medium is diluted to quarter-strength. The diluted medium is solidified with 0.45% (w/v) Gelrite. The pH is adjusted to 5.8 ± 0.1

prior to autoclaving. The medium is poured into sterile Petri dishes. The original SH medium was described by Schenk and Hildebrandt *(19)*.

2.2. Cryopreservation

2.2.1. Sorbitol Treatment

1. 4 M sorbitol solution.

2.2.2. Dimethyl-sulfoxide treatment

1. Dimethyl-sulfoxide (DMSO).

2.2.3. Freezing procedure

1. Programmable freezer.

2.3. Gene Transfer

2.3.1. Pre-Treatment

1. Solidified half-strength LP medium (*see* **Subheading 2.1.1**) containing 0.25 M *myo*-inositol.

2.3.2. Bombardment

1. To coat gold particles with DNA, to 10 mg of gold particles (1.5–3.0 μm in diameter) suspended in 210 μL water in an Eppendorf tube is added 20 μg of plasmid DNA in 10 mM Tris–1 mM EDTA buffer, pH 8.0, with vortexing followed by one-tenth volume 3 M sodium acetate and 2.5 volumes ethanol with vortexing at each step. The suspension is placed at –20 °C for 30 min, and centrifuged very briefly at 3000 g, after which the supernatant is removed and replaced with 1 mL ethanol.
2. Equipment for bombardment.

2.3.3. Selection of Transformed Cells

1. Modified half-strength LP medium (*see* **Subheading 2.1.1**) containing 0.25 M *myo*-inositol and with glutamine replaced by 1 mM asparagine, solidified with 0.18% (w/v) Gelrite.
2. Half-strength LP medium as under 1 to which Basta (1 mg/L) has been added. Basta is a commercial preparation (Hoechst Ltd) of ammonium glufosinate, 200 g/L. It is incorporated, filter-sterilized in the culture medium at 1 mg Basta solution per liter, corresponding to 0.2 mg/L ammonium glufosinate.
3. Half-strength solidified LP medium (*see* **Subheading 2.1.1**)

3. Methods

3.1. The System of Somatic Embryogenesis in Norway Spruce

The whole process of somatic embryogenesis in Norway spruce can be divided into the following steps: initiation, proliferation, pre-maturation, maturation, partial desiccation and germination.

3.1.1. Initiation

The primary explant for establishing embryogenic cultures is crucial for obtaining a high initiation frequency. About 80% of the embryos will give rise to embryogenic cultures when using fully mature but not desiccated zygotic embryos. Embryogenic cells differentiate from the hypocotyl and/or cotyledons *(20)*. The embryogenic cells originate from nodules that are formed either from epidermal, subepidermal or cortical cells. Embryogenic cultures are initiated within 1 month. It is fairly easy to distinguish between embryogenic and non-embryogenic cultures as the latter are mucilagenous and translucent and consist of PEMs. Embryogenic cultures initiated from single explants are kept as separate lines. Consequently, each line represents one genotype. The cultures are regarded as established when the tissue increases in size at least two times within a subculture interval. Embryogenic cell lines often lose their potential to differentiate somatic embryos after continuous proliferation for several months. The risk for this degenerative process is especially high when the cultures are proliferating in liquid medium. In order to avoid this problem, it is advisable to keep all embryogenic lines cryopreserved in liquid nitrogen and thaw the cultures 3–5 months before start of the experiments. As soon as an embryogenic culture is established, it should be cryopreserved in liquid nitrogen for future use (*see* **Note 1**).

1. Seeds are surface-sterilized in 70% ethanol for 5 min and then rinsed three times in sterile water. The seeds are finally soaked in sterile water and stored at 4 °C for 12–24 h.
2. Isolated zygotic embryos are inoculated on the initiation medium (*see* **Subheading 2.1.1**) at 20 °C in darkness. Different culture media can be used for initiation of embryogenic cultures, but we have found that half-strength modified LP medium gives stable results.
3. The explants are subcultured to fresh medium every 3–4 weeks. New cell lines are established after 3–5 months.
4. Cryopreservation of newly established cell lines (*see* **3.2**).

3.1.2. Proliferation

Proliferation of embryogenic cultures takes place under the same conditions as initiation. However, the developmental processes can only be synchronized provided that proliferation takes place in liquid medium. In suspension cultures, single cells and cell aggregates develop as separate structures. Thus, the cells can easily be separated by sieving or centrifugation, and thereafter, subcultured and maintained as required. As long as auxin and cytokinin are present in the culture medium, unequal division of embryogenic cells with dense cytoplasm leading to the restart of the process from PEM I level continues. It should be noted that auxin is rapidly taken up from the medium so that depletion of auxin starts already after a few days. Consequently, if the cultures are not frequently transferred to fresh culture medium, the development of somatic embryos will start.

It has long been known that conditioned medium from embryogenic cultures can promote embryogenesis. The ability of conditioned medium to sustain or stimulate somatic embryogenesis implies that secreted soluble signal molecules play an important role. Several components in conditioned medium have been found to promote somatic embryogenesis. In Norway spruce, we have found that extracellular chitinases *(21)*, arabinogalactan proteins *(22)* and lipo-chitooligosaccharides *(23)* affect somatic embryogenesis. The amount of secreted factors varies among cell lines and also depends on the developmental stage of PEMs and somatic embryos. Because embryogenic cells secrete conditional factors that affect proliferation rate and differentiation of somatic embryos, it is important to strictly control the density of the cultures.

1. Embryogenic cultures are proliferated on solidified medium under the same conditions as during initiation. The cultures are transferred to fresh medium every 3–4 weeks.
2. Suspension cultures are established 2 months prior to the onset of the experiments (*see* **Note 2**).
3. Embryogenic cultures are maintained by weekly or biweekly subculturing 3 mL of settled cell aggregates into 97 mL proliferation medium (in 250-mL Erlenmeyer flasks) containing half-strength LP medium as described for initiation, but without Gelrite. Suspension cultures are grown on a gyratory shaker (100 rpm) in darkness at 20 °C.

3.1.3. Pre-Maturation

Comprising a link between the proliferation of PEMs and the highly organized embryo development, PEM to embryo transition plays a crucial role in somatic embryogenesis. PEM to embryo transition occurs after withdrawal of auxin and cytokinin. The inability of a cell line to form cotyledonary somatic embryos is to a large extent associated with disturbed or arrested PEM to embryo transition. The embryos should not be exposed to maturation treatments before they have reached the developmental stage corresponding to early embryogeny. The time for the transition from PEMs to somatic embryos varies for different lines, but usually the switch from proliferation to development occurs after 24 h *(24)*. Newly formed somatic embryos should develop in medium lacking auxin and cytokinin within 1 week. After this period, they have reached the stage comparable to early embryogeny of zygotic embryos and obtained the ability to respond to ABA.

When cultures are transferred from proliferation medium to medium stimulating development, they consist of a mixture of single cells and cell aggregates. In order to synchronize development, the cells can be washed.

1. Three millilitres of settled cell aggregates collected from 1-week-old suspension cultures are successively washed in 15-mL Falcon tubes with two 10 mL aliquots of half-strenth LP medium lacking 2,4-D and BA.
2. Three millilitres washed, supernatant-free cell aggregates are inoculated into 97 mL half-strength LP medium lacking 2,4-D and BA in 250-mL Erlemmeyer flasks. The cultures are grown on a gyratory shaker (100 rpm) at 20 °C in darkness for about 1 week until early somatic embryos have differentiated (*see* **Note 3**).

3.1.4. Maturation

Once early somatic embryos have been formed, their further development to mature forms requires ABA. The yield of cotyledonary somatic embryos is dependent on the presence of early somatic embryos in the suspension culture at the time when it is subjected to ABA treatment. The root-organizing centre is established at the beginning of late embryogeny, after about 2 weeks. After 5 weeks, the somatic embryos have developed provascular tissue, cortex, root and shoot meristems. When the embryos have reached this stage, they should be exposed to partial desiccation. Prolonged contact with ABA during somatic embryo maturation has a negative influence on later plant growth *(17)*.

1. Two mililitres of settled cell aggregates, corresponding to 300 mg fresh weight biomass, is plated onto Whatman no. 2 filter papers placed in Petri dishes on solidified BMI-SI medium (*see* **Note 4**). The cultures are incubated in darkness at 20 ± 1 °C.
2. Filter papers with thin layers of maturing somatic embryos are transferred to fresh medium every second week.

3.1.5. Partial Desiccation and Germination

Maturation of zygotic embryos of Norway spruce is generally concluded with some degree of drying, which results in a gradual reduction of metabolism as water is lost from the seed tissue and the embryo passes into a metabolically inactive, or quiescent, state. Hydration of the seed leads to its germination, which is an immediate response to a switch from the maturation programme to the germination programme. In order to avoid precocious germination of somatic embryos, the embryos should be given a desiccation treatment. During the desiccation treatment, the water content of the embryos decrease to 25–35% within the first 8 days and does not change after prolonged treatment *(17)*. Partially desiccated embryos are germinated (*see* **Note 5**).

1. For partial desiccation, cotyledonary somatic embryos are transferred to empty 35-mm Petri dishes placed inside a 90-mm Petri dish containing 10 mL sterile water. The large Petri dish is sealed with Parafilm and left for 8–24 days in darkness at 20 °C.
2. Germination is carried out on quarter-strength solidified SH medium (*see* **Subheading 2.1.5**). Cultures are kept in the dark for the first 4 days followed by incubation under continuous light (120 µmol/m^2/s) from fluorescent tubes (Philips 58W/84) supplemented with incandescent light at 20 ± 1 °C. Germination takes 5 weeks without subculture.

3.2. Cryopreservation

Fast growing embryogenic cells can be cryopreserved (*see* **Note 6**). Embryogenic cultures of Norway spruce are cryopreserved according to a protocol adjusted by Nörgaard et al. *(25)*.

3.2.1. Sorbitol Treatment

1. Day 1: About 3 g of fast proliferating embryogenic tissue, 1 week after subculture, is transferred to 40 mL half-strength LP liquid medium lacking auxin and cytokinin (*see* **Subheading 2.1.3**). Two millilitres of a 4 M sorbitol solution is added in small parts for 30 min. The suspension culture is kept on a rotary shaker at 100 rpm for 24 h.
2. Day 2: Sorbitol (4 M solution) is added to the suspension to give a final concentration of 0.4 M sorbitol, and the cultures are placed on a rotary shaker for another 24 h.

3.2.2. Dimethyl-Sulfoxide Cryoprotective Treatment

1. The culture vials are transferred onto ice for approximatively 15 min.
2. DMSO is added to the suspension in small amounts for 30 min to give a final concentration of 5% DMSO.
3. Settled cell aggregates, 1.0–1.5 mL, is dispersed into 2-mL cryovials and placed on ice before freezing.

3.2.3. Freezing Procedure

1. The embryogenic cultures are frozen in a programmable controlled-temperature cooling chamber. The embryogenic tissue is frozen at a rate of 0.3 °C/min to the temperature of –16 °C and held at –16 °C for 15 min. Thereafter, the material is frozen at a rate of 0.3 °C/min to the terminal temperature of –35.5 °C.

2. After reaching the terminal temperature, the cryotubes are immediately immersed in liquid nitrogen. The tubes are stored in liquid nitrogen.

3.2.4. Thawing

1. The cryotubes are thawed quickly in sterile water at 45 °C (*see* **Note 7**).
2. Thereafter, the tubes are first transferred to sterile water at 4 °C for 2–3 min and then to 70% ethanol for 2–3 min.
3. The cell aggregates in the tubes are poured out onto a sterile filter paper placed in Petri dishes on solidified proliferation medium.
4. The filter papers are transferred to fresh proliferation medium after 1, 18 and 24 h.
5. Thereafter, the filter paper with thawed tissue is transferred to fresh proliferation medium every second week.
6. Regrowth starts after 2–4 weeks. When the colonies of proliferating embryogenic tissue have reached a size of 2–3 mm in diameter, they are isolated and cultured individually.

3.3. Gene Transfer

Particle bombardment is usually the best method for delivering DNA to cells for assays of promoter activity by transient expression. Furthermore, particle bombardment is routinely used for producing transgenic plants of Norway spruce. Best results are obtained with cell lines that have been in culture for about 2–3 months after thawing (*see* **Note 8**).

To compare the activity of various promoters, the most efficient procedure is to incorporate an internal control *(26)*. For example, to coat the gold particles with both a plasmid containing a promoter-*gusA* construct and a control plasmid containing the *luc* gene coding for luciferase under a 35S promoter. The use of the internal control enables a correction to be made for shot-to-shot variation and reduces the amount of replication required to detect significant differences in promoter activity.

The method used for producing transgenic embryos of Norway spruce was originally published by Clapham et al. *(27,28)*.

3.3.1. Pre-Treatment

1. The cultures are maintained by weekly subculture in proliferation medium (*see* **Subheading 2.1.2., step 1**). Four to five days after subculture, cells are allowed to sediment in 50-mL Falcon tubes and are plasmolyzed by resuspension in proliferation medium containing *myo*-inositol and asparagine instead of glutamine (*see* **Subheading 2.3.3., step 2**). Cell samples of 100–150 mg fresh weight are collected by vaccum filtration on 4 cm diameter circles of filter paper

(Munktell 1F). These are placed on a bed of four 12-cm diameter filter papers soaked in the same medium, in 15-cm Petri dishes, until bombardment 1–3 h later.

3.3.2. Bombardment

1. The cells are bombarded with gold particles coated with DNA. Various types of apparatus for bombardment can be used; we employ a modified particle inflow gun.

3.3.3. Selection for Transformed Cells

1. The filter paper discs with bombarded cells are transferred to solidified half-strength LP medium containing *myo*-inositol and asparagine (*see* **Subheading 2.3.3., step 1**) directly after bombardment (*see* **Note 9**).
2. After 8 days, the filter papers are transferred to fresh proliferation medium to which Basta has been added (*see* **Subheading 2.3.3., step 2**). They are subcultured on this medium monthly. Embryogenic tissue resistant to Basta appears from 2–4 months after bombardment.
3. When colonies of proliferating embryogenic tissue are about 2 mm in diameter, they are isolated and subcultured to standard proliferation medium (*see* **Subheading 2.1.2**) giving rise to new transgenic lines within 1–2 months.
4. DNA and RNA are isolated from samples of 100–150 mg from putative transformants and tested by polymerase chain reaction for the presence of the transgene.

3.3.4. Regeneration of Transgenic Plants

1. The transgenic lines are treated as untransfomed embryogenic cultures as described in **Subheadings 3.1.2.** to **3.1.5.**

4. Notes

1. When establishing suspension cultures, it is crucial that the cell and cell aggregate density is kept high. If the cultures are growing slowly, old medium should be removed before new medium is added. If limited amounts of the culture are available when suspension cultures are to be established, it is better to use small Erlenmayer flasks (25 or 100 mL).
2. Embryogenic cultures successively lose their embryogenic potential after prolonged continuous proliferation. When the embryogenic potential has decreased, it can be difficult to synchronize the cultures. Most cell lines are best suited for developmental studies 3–5 months after thawing.
3. PEM to somatic embryo transition occurs within a short period after withdrawal of auxin and cytokinin. The pre-treatment should be given for only a limited period, usually 1 week. After prolonged treatment, the early somatic embryos degenerate. ABA is incapable of inducing PEM to somatic embryo transition.

This may be responsible for the failure to stop proliferation when embryogenic cultures are directly transferred from medium containing auxin and cytokinin to maturation medium.

4. Polyethylene glycol (PEG) with a molecular weight greater than 4000 is often added to the maturation medium in order to increase the frequency of mature somatic embryos. However, for Norway spruce, PEG treatment exerts a deleterious effect on embryo morphology and root meristem development. By excluding PEG from the maturation medium, the yield of cotyledonary embryos decreases but their quality increases.

5. The sensitivity of the partially desiccated embryos to stress imposed by rehydration (imbibition) is controlled by the initial moisture content of the embryo and the rate at which water is taken up. The interaction of these factors has a dramatic effect on germination. It is important to keep the initial rate of water uptake at a low level.

6. When cryopreserving embryogenic cultures, it is crucial that they are healthy and fast growing. Both cultures grown on solidified medium and in liquid medium can be cryopreserved. However, if cultures grown on solidified medium are used, there is risk that the proliferation rate is not high enough, resulting in the presence of vacuolated cells. Only meristematic cells will survive the freezing procedure and gain a high regrowth after thawing.

7. The tubes should be moved from 45 °C to 4 °C as soon as the cell aggregates begin to melt.

8. After 6 months in culture, the transformability is still high but the quality of the PEMs deteriorates so that it can be harder to regenerate plants from the transformed cells.

9. Asparagine replaces glutamine in the media for transformation because glutamine interferes with selection on glufosinate, an inhibitor of glutamine synthase. The incorporation of 5-azacytidine into selection media for the first week, as recommended earlier, seems to be unnecessary.

References

1. Singh H. Embryology of gymnosperms, In: Zimmerman W, Carlquist Z, Ozenda P, Wullf HD, eds. Handbuch der Pflanzen anatomie. Berlin: Gebryder Borntraeger, 1978:187–241.

2. Filonova LH, Bozhkov PV, von Arnold S. Developmental pathway of somatic embryogenesis in *Picea abies* as revealed by time-laps tracking. J Exp Bot 2000;343:249–264.

3. Van Zyl L, Bozhkov PV, Clapham D, Sederoff R, von Arnold S. Up, down and up again is a signature global gene expression pattern at the beginning of gymnosperm embryogenesis. Gene Expr Patterns 2003;3:83–91.

4. Stasolla C, Bozhkov PV, Chu T-M, van Zyl L, Egertsdotter U, Suárez MF, Craigh D, Wolfinger R, von Arnold S. Sederoff R. Variation in transcript abundance during somatic embryogenesis in gymnosperms. Tree Physiol 2004;24:1073–1085.

5. Filonova LH, Bozhkov PV, Brukhin V, Danierl G, Zhivotovsky B, von Arnold S. Developmental programmed cell death in plant embryogenesis: exploring a model system of Norway spruce somatic embryogenesis. J Cell Sci 2000; 113: 4399–4411.

6. Bozhkov PV, Filonova LH, Suárez MF, Helmersson A, Smertenko AP, Zhivotovsky B, von Arnold S. VEIDase is a principal caspase-like activity involved in plant programmed cell death and essential for embryonic pattern formation. Cell Death Differ 2004;11:175–182.

7. Bozhkov PV, Suárez MF, Filonova LH, Daniel G, Zamyatnin AA, Rodriguez-Nieto S, Zhivotovsky B, Smertenko AP. Cysteine protease mcII-Pa executes programmed cell death during plant embryogenesis. Proc Natl Acad Sci USA 2005;102:14463–14468.

8. Suárez MF, Filonova L, Smertenko AP, Savenkov E, Clapham C, von Arnold S, Zhivotovsky B, Bozhkov PV. Metacaspase-dependent programmed cell death is essential for plant embryogenesis. Curr Biol 2004;14:R339–R340.

9. Smertenko AP, Bozhkov PV, Filonova LH, von Arnold S Hussey P. Re-organisation of the cytoskeleton during developmental programmed cell death in *Picea abies* embryos. Plant J 2003;33:813–824.

10. Ingouff M, Farbos I, Lagercrantz U, von Arnold, S. *PaHB1* is an evolutionary conserved HD-GL2 homeobox gene defining the protoderm during Norway spruce embryo development. Genesis 2001;30:220–230.

11. Sabala I, Clapham D, Elfstrand M, von Arnold S. Tissue-specific experssion of Pa18, a putative lipid transfer protein gene, during embryo development in Norway spruce (*Picea abies*). Plant Mol Biol 2000;42:461–478.

12. Hjortswang H, Sundås A, Bharathan G, Bozhkov PV, von Arnold S, Vahala T. *KNOTTED 1*-like homeobox genes of a gymnosperm, Norway spruce, expressed during somatic embryogenesis. Plant Physiol Biochem 2002;40:837–843.

13. Ciavatta V, Egertsdotter U, Clapham C, von Arnold S, Cairney J. A promoter from the loblolly pine *PtNIP1;1* gene directs expression in an early-embryogenesis and suspensor-specific fashion. Planta 2002;215:694–698.

14. Ingouff M, Farbos M, Wiweger M, von Arnold S. The molecular characterization of *PaHB2*, a homeobox gene of the HD-GL2 family expressed during embryo development in Norway spruce. J Exp Bot 2003;54:1343–1350.

15. Footitt S, Ingouff M, Clapham D, von Arnold S. Expression of the viviparous 1(*Pavp1*) and p32 protein kinase (*cdc2Pa*) genes during somatic embryogenesis in Norway spruce (*Picea abies*). J Exp Bot 2003;54:1711–1719.

16. von Arnold S, Eriksson T. A revised medium for growth of pea mesophyll proto-plasts. Plant Physiol 1977;39:257–260.

17. Bozhkov PV, von Arnold S. Polyethylene glycol promotes maturation but inhibits further development of *Picea abies* somatic embryos. Physiol Plant 1998;104: 211–224.

18. Krogstrup P. Embryo-like structures from cotyledons and ripe embryos of Norway spruce (*Picea abies*). Can J For Res 1986;16:164–168.

19. Schenk U, Hildebrandt AC. Medium and techniques for induction and growth of monocotyledonous and dicotyledonous plant cell cultures. Can J Bot 1972;50: 199–204.

20. Mo LH, von Arnold S. Origin and development of embryogenic cultures from seedlings of Norway spruce (Picea abies). J Plant Physiol 1991;138:223–230.

21. Wiweger M, Farbos I, Ingouff M, Lagercrantz U, von Arnold, S. Expression of *Chia4-Pa* chitinase genes during somatic and zygotic embryo development in Norway spruce (*Picea abies*): similarities and differences between gymnosperm and angiosperm class IV chitinases. J Exp Bot 2003;54:2691–2699.

22. Egertsdotter U, von Arnold S. Importance of arabinogalactan proteins for the development of somatic embryos of *Picea abies*. Physiol Plant 1995;93:334–345.

23. Dyachok J, Wiwegwe M, Kenne L, von Arnold, S. Endogenous Nod-factor-like signal molecules suppress cell death and promote early somatic embryo development in Norway spruce. Plant Physiol 2002;128:523–533.

24. Bozhkov PV, Filonova LH, von Arnold S. A key developmental switch during Norway spruce somatic embryogenesis is induced by withdrawal of growth regulators and is associated with cell death and extracellular acidification. Biotechnol Bioeng 2002;**77**:658–667.

25. Nörgaard JV, Duran V, Johnsen Ö, Krogstrup P, Baldursson S, von Arnold S. Variations in cryotolerance of embryogenic *Pice abies* cell lines and association to genetic, morphological and physiological factors. Can J For Res 1993;23: 2560–2567.

26. Clapham DH, Häggman H, Elfstrand M, Aronen T, von Arnold S. Transformation of Norway spruce (*Picea abies*) by particle bombardment. In: Jackson JF, Linskens HF, Inman RB, eds. Molecular Methods of Plant Analysis. Berlin Heidelberg Springer-Verlag, vol. 23, 2003:127–146.

27. Clapham D, Demel P, Elfstrand M, Koop H-U, Sabala I, von Arnold S. Effective biolistic gene transfer to embryogenic cultures of *Picea abies* and the production of transgenic plantlets. Scand J For Res 2000;15:151–160.

28. Clapham D, Newton R, Sen S, von Arnold, S. Transformation of *Picea* species. In: Jain M, Minocha SC eds. Molecular Biology of Woody Plants. Dordrecht: Kluwer, vol. 2. 2000:105–118.

II

CELLULAR, GENETIC AND MOLECULAR MECHANISMS OF PLANT EMBRYOGENESIS

4

In Vitro Fertilization With Isolated Higher Plant Gametes

Erhard Kranz, Yoichiro Hoshino, and Takashi Okamoto

Summary

Methods have been developed to isolate gametes of higher plants and to fertilize them in vitro. Zygotes, embryos, fertile plants and endosperm can now be obtained from in vitro fusion of pairs of sperm and egg cells and of pairs of sperm and central cells, respectively. This allows examination of the earliest developmental processes precisely timed after fertilization. The isolated egg and central cell, fertilized and cultured in vitro, are able to self-organize apart from each other and without mother tissue in the typically manner. Thus, this system is a powerful and unique model for studies of early zygotic embryogenesis and endosperm development. The underlying processes are now comparatively studied in detail by investigations of expression of genes and their corresponding proteins. The use of these techniques opens new avenues in fundamental and applied research in the areas of developmental and reproductive plant biology.

Key Words: In vitro fertilization; egg cells; zygotes; endosperm; micromanipulation; single cell analysis; maize.

1. Introduction

The application of several micromanipulation techniques enables the isolation and in vitro fusion of female and male angiosperm gametes. Furthermore, single cell culture enables development of single or a few zygotes and endosperm. Thus, in vitro fertilization (IVF) includes the combination of three basic microtechniques: (i) the isolation, handling and selection of male and female gametes, (ii) the fusion of pairs of gametes and (iii) the single cell culture. This experimental system enables studies of gamete interaction and hybridization as well as cytological and molecular events, which occur immediately after gamete fusion. In maize, starting with gamete fusion, the development of a

From: *Methods in Molecular Biology, vol. 427: Plant Embryogenesis*
Edited by: M. F. Suárez and P. V. Bozhkov © Humana Press, Totowa, NJ

single zygote into an embryo and finally into a plant and the development of an in vitro fertilized central cell into endosperm can be followed in vitro *(1,2)*.

In contrast to animal and lower plants, IVF with angiosperms presupposes the isolation of gametes. Egg and central cells have to be isolated from an embryo sac which is generally embedded in the nucellar tissue of the ovule and normally contains two synergids and some antipodal cells. Moreover, sperm cells have to be isolated from pollen grains or tubes. Double fertilization, that is the fusion of one sperm with the egg to create the embryo and the fusion of the other sperm with the central cell to form the endosperm *(3)*, occurs in vivo deeply within ovule tissues in the embryo sac and generally with the help of one of the two synergids.

Female gametes of, for example barley, wheat and rape seed can be generally isolated by mechanical means *(4–7)* but also by using mixtures of cell wall degrading enzymes in combination with a manual isolation procedure. In maize, the same can be achieved *(8)*, but it is more effective to treat the nucellar tissue containing the embryo sac with such enzymes for a short period prior to the manual isolation step in order to soften this tissue and to avoid damage of gametic protoplasts. Isolated sperm cells are obtained from pollen grains (tricellular pollen) or tubes (bicellular pollen) by osmotic bursting, squashing or grinding of the material.

Because they are protoplasts, individual isolated gametes have been fused electrically (e.g., *see* **refs.** *(1,2,8–12)* and chemically, by calcium *(13–15)* or by polyethyleneglycol *(16)*. Those fusiogenic media including calcium might be used to determine conditions and factors, which promote adhesion, in vivo membrane fusion and possibly recognition events taking place during the fertilization process. This method was used to study differential contribution of cytoplasmic Ca^{2+} and Ca^{2+} influx to gamete fusion and egg activation *(17)*. The fusion method using electrical pulses for cell fusion is described here in detail because it is well established and an efficient part of the IVF procedure.

Very early steps in zygote and endosperm development, precisely timed after gamete fusion, can be analyzed in microdroplet culture under defined conditions without feeder cells *(8,12,18)*. To achieve sustained growth of zygotes and endosperm, cocultivation with feeder cells is suitable *(1,2,4,8,11,19)*. Embryogenesis, plant regeneration and endosperm development from isolated male and female gametes fused in vitro have, so far, been achieved exclusively in maize using electrofusion techniques and nurse culture *(1,2)*. In this system, zygote, embryo, plant and endosperm development takes place in the absence of mother tissue, as is the case with endosperm formation without an embryo and embryo development occurs without endosperm in a self-organizing manner. Therefore, all IVF procedures with isolated, higher plant gametes are different from that in vivo.

With microtechniques originally developed for somatic protoplasts *(20,21)*, defined gamete fusion is possible. The possibility of selection, individual transfer and handling of single gametes, zygotes and primary endosperm cells together with a high frequency of fusion and cell divisions enables physiological and molecular studies at the single cell level. For example, by using single-cell micromanipulation techniques, we developed an immunocytochemical procedure to examine subcellular protein localization in isolated and cultured single cells *(22)*. This method is described in this chapter. Because of the efficient production of in vitro zygotes by electrofusion, sufficient numbers of such cells allows molecular analyses for gene and protein expression studies of genes involved in early events occurring during zygote formation, early embryogenesis and endosperm development *(19,23–27)*. By use of reverse transcription-polymerase chain reaction (RT-PCR) methods, cDNA libraries have been generated from egg cells *(28)* and in vitro zygotes *(29)* to isolate egg- and central cell-specific *(30)* and fertilization-induced genes (e.g., *see* **refs**. *23,24*).

We describe tools of IVF and micromanipulation techniques, for example, to separate apical and basal cells from the two-celled embryo and gene isolation methods for studies of cell-specific gene expression in these cells to elucidate mechanisms of early embryonic patterning in higher plants *(23,24, see* **Fig. 3**). Moreover, a description of an adapted method for the analyses of lysates from a few egg cells and zygotes by polyacrylamide gel electrophoresis and subsequent mass spectrometry-based proteomics technology is given to identify major protein components expressed in these cells *(25,26, see* **Fig. 3**). Unless otherwise mentioned, the procedures are described for maize *(Zea mays)*. Further potentials of IVF for studies on mechanisms of fertilization and early development have been reviewed (e.g., *see* **refs**. *31–33*), and micromanipulation techniques have been described elsewhere *(34,35)*.

2. Materials

2.1. Isolation and Transfer of Gametes

1. Plants grown in the greenhouse under standard conditions (*see* **Note 1**).
2. Laminar flow box.
3. Inverted microscope.
4. Plastic dishes, 3 cm.
5. Cover slips, siliconized at the edges with Repel-Silane (Merck; Pharmacia Biotech) and UV-sterilized.
6. Mineral oil, autoclaved (Paraffin liquid for spectroscopy, Merck).
7. Mannitol solution adjusted to 570 mosmol/kg H_2O before autoclaving (concentration about 530 mM), after autoclaving 600 mosmol/kg H_2O.
8. Mixture of cell wall degrading enzymes: 1.5% pectinase (Serva), 0.5% pectolyase Y23 (Seishin, Tokyo), 1% cellulase Onozuka RS (Yakult Honsha)

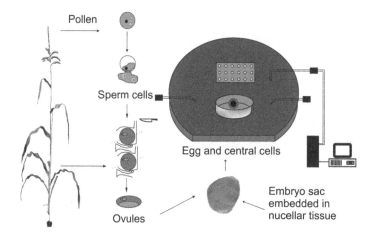

Fig. 1. Set up for in vitro fertilization experiments in maize: isolation, selection, transfer and fusion of single gametes. Ear spikelets are cut as indicated (dotted lines). After removing integuments, cells of the embryo sac are isolated from nucellar tissue pieces in the isolation chamber (plastic dish) manually with needles following transfer with a capillary, which is connected via a hydraulic system to a computer-directed micropump into microdroplets on a cover slip. Subsequently, sperm cells are selected in the isolation chamber after release from the pollen grains by osmotic shock and transferred into microdroplets for gamete fusion.

and 1% hemicellulase (Sigma) dissolved in bidistilled water, adjusted to 570 mosmol/kg H_2O with mannitol and pH 5.0, and filter-sterilized before freezing. Store 10 mL aliquots at –20 °C.
9. Sliding stage for the insertion of a cover slip and a plastic dish, self-made (*see* **Fig. 1**).
10. Glass needles with fine tips, preferably prepared using a microforge.
11. Capillaries, tip openings 100–300 μm (drawn by hand) and 20 μm (drawn by a puller).
12. Cell transfer Systems: Computer-controlled Dispenser/Dilutor (Microlab-M, Hamilton, or Nano Spuit, IKEDA Scientific, or CellTram, Eppendorf).

2.2. Fusion of Gametes

1. Electrofusion apparatus (CFA 500, Krüss).
2. Electrodes and electrode support, self-made.
3. Positioning system (optional), especially useful for gently moving the electrodes along the z-axis by a step motor (type MCL, Lang).
4. Electrodes (platinum wire, diameter 50 μm) fixed to an electrode-support, which is mounted under the condenser of the microscope.

2.3. Culture of Zygotes and Central Cells, Embryogenesis and Regeneration

1. Plastic dishes, 3 cm.
2. Four-well multidish (Nunclon; Nunc A/S).
3. Feeder cells: cereal suspension cells (*see* **Note 2**).
4. Zygote Murashige Skoog medium (ZMS) for in vitro zygotes and feeder cells: Murashige Skoog medium (MS) *(36)* with the modifications of NH_4NO_3 (165 mg/L) and organic constituents. These are nicotinic acid (1 mg/L), thiamine HCl (10 mg/L), pyridoxine HCl (1 mg/L), L-glutamine (750 mg/L), proline (150 mg/L), asparagine (100 mg/L), myo-inositol (100 mg/L) and 2,4-D (2 mg/L), adjusted to 600 mosmol/kg H_2O with glucose, pH 5.5 and filter-sterilized.
5. "Millicell-CM" inserts, diameter 12 mm (Millipore).
6. Regeneration media (RMS): Filter-sterilized MS medium *(36)* solidified with 4 g/L agarose (type I-A; Sigma) and its modifications: RMS1, medium without hormones and supplemented with 60 g/L sucrose; RMS2, same medium as RMS1 but containing 40 g/L sucrose; RMS3, medium without hormones and supplemented with 10 g/L sucrose, macrosalts and microsalts half concentration.

2.4. Identification of Proteins

1. Sodium dodecyl suphfate (SDS) sample buffer: 2% SDS, 25 mM Tris–HCl (pH 6.8), 30% glycerol, 5% 2-mercaptoethanol.
2. Lysis buffer for isoelectric focusing: 8 M urea, 5% 2-mercaptoethanol, 2% (v:v) Ampholine (pH 3.5–10) (Amersham), 2% Nonidet P40.
3. Isoelectric focusing gel: 8 M urea, 3.5% acrylamide, 0.18% Bis-acrylamide, 5% (v:v) ampholine (pH 3.5–10), 2% Nonidet P40.
4. Compact slab gel electrophoresis system (AE7300 system, ATTO).
5. Isoelectric focusing gel electrophoresis system with which thin glass capillary (length, 5 cm; diameter, 1 mm) should fit.

2.5. Identification and Expression of Genes

1. Dynabeads mRNA DIRECT™ Micro Kit (Dynal Biotech).
2. Super SMART™ PCR cDNA Synthesis Kit (BD Biosciences).
3. Randomly amplified polymorphic DNA (RAPD) primers: 10-mer DNA primers from Operon (Kits A, B, O and T) and the University of British Colombia (Set 100/1).
4. Thermal cycler for PCR and cDNA synthesis.

2.6. Immunocytochemical Techniques

1. Fixative solution: 4% (w/v) paraformaldehyde, 0.1% (v/v) glutaraldehyde and 600 mosmol mannitol in a microtubule-stabilizing buffer (MSB) at pH 6.9 *(37,38)*, with minor modifications, containing 50 mM 1,4-piperazinediethanesulfonic acid, 5 mM ethylene glycol bis (2-aminoethyl ether)-N,N,N´,N´-tetraacid, 1 mM $MgCl_2$, and 2% (v/v) glycerol).

2. Poly-L-lysine: 0.1% (w/v) in water (Sigma).
3. Adhesive tapes.
4. Anti-α-tubulin monoclonal antibody (DM1A; NeoMarkers) diluted 1:100 in MSB.
5. Bovine serum albumin (BSA), 0.25% (w/v).
6. Fluorescein isothiocyanate (FITC)-conjugated anti-mouse IgG antibody raised in goat (Sigma).
7. 4′,6-diamidino-2-phenylindole (DAPI, 0.1 µg/mL in phosphate-buffered saline).
8. Antifading solution containing 100 mg/L 1,4-diazabicyclo [2,2,2] octane (Sigma).

3. Methods

3.1. Isolation of Gametes

1. Collect pollen in the morning from freshly dehisced anthers. Immediately after pollen collection, store pollen for several hours at room temperature in plastic dishes containing a wet filter paper fixed to the lid to provide a moistened atmosphere.
2. After silk emergence, collect ears and sterilize the outer leaves with ethanol (70%).
3. Dissect 20–30 nucellar tissue pieces from the ovules under a dissecting microscope (*see* **Fig. 1**). The embryo sac should be visible in the nucellar tissue pieces. Collect the tissue pieces in 1 mL mannitol solution (600 mosmol/kg H_2O) in 3-cm plastic dishes and add 0.5 mL enzyme mixture (*see* **Note 3**). Incubate the mixture at room temperature for 30 min without shaking. After treatment, the dishes can be stored in the refrigerator at 6°C (*see* **Note 4**).
4. Overlay the siliconized and UV-sterilized cover slip with 300 µL autoclaved mineral oil and inject 2 µL mannitol droplets (600 mosmol/kg H_2O) in three rows, each with 10 droplets using a microcapillary and a micropump. Take care that the droplets do not spread over the glass surface, but are located inside the oil and have no access to the air.
5. Isolate manually egg cells and the other cells of the embryo sac directly in the incubation dish with glass needles under microscopic observation and transfer the cells by a microcapillary (tip opening 100–200 µm for egg cells and 300 µm for central cells) into the microdroplets using a micropump (*see* **Notes 1** and **4**; **Fig. 1**).
6. For central cell isolation, select ovules from the middle part of ears (emerged silk length, 3–18 cm; ovule diameter, 2–3 mm), which were bagged before silk emergence. In contrast to egg cell isolation, only the nucellar cells at the micropylar end of the embryo sac are removed from the tissue pieces. Starting at the chalazal end near the antipodal cells, the central cell is then pushed by a microneedle toward the micropylar end of the embryo sac, where it is liberated and becomes spherical. Approximately 20–30 tissue pieces containing the embryo sac are collected in a 1 mL mannitol solution (osmolality of 750 mosmol/kg H_2O) in 3-cm diameter plastic dishes, followed by the addition of 0.5 mL mannitol

solution (570 mosmol/kg H_2O) containing the enzyme mixture. Incubate at room temperature for 45 min to 1 h without shaking.

7. Overlay about 1000 pollen grains in a 3.5-cm diameter plastic dish with 1.5 mL mannitol solution (600 mosmol/kg/H_2O). After grain bursting, transfer the sperm cells (tip opening of the capillary 20 μm) by use of a micropump into the microdroplets containing egg or central cells (*see* **Notes 5** and **6; Fig. 1**).

3.2. Fusion of Gametes

1. Mount an electrode support with two fixed electrodes (50-μm diameter platinum wire) under the condenser of the microscope. Before use, sterilize the ends of the electrodes in a weak flame. Adjust the electrodes to the crosshairs position and lower these onto the cover slip and into one mannitol droplet.

2. Prepare and adjust the electrodes carefully as demonstrated in **Fig. 2**. Align and fix the two gametes at one electrode. By moving the microscope stage, first move one egg or central cell towards one electrode. Finally, the egg or central cell is fixed to the electrode by dielectrophoresis (1 MHz, 70 V/cm). Using the same procedure, the sperm cell is fixed to the female gamete. The electrical conditions for dielectrophoretic alignment of a central and a sperm cell are 1 MHz and 38–56 V/cm. Now the final distance of the electrodes is adjusted. It should be approximately twice the sum of the diameters of the cells.

3. Induce cell fusion by applying a single or a maximum of three negative DC-pulses (50 μs; 0.9–1.0 kV/cm for sperm–egg fusion; *see* **Note 7**). Central cell–sperm fusion is induced by a single or by two to three negative direct current pulses (50 μs; 0.4–0.5 kV/cm).

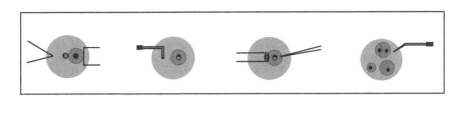

Fig. 2. Shown from left to right are manipulations in microdroplets placed on a cover slip for cell fusion, staining and immunochemistry, microinjection and cell separation for further analysis and culture.

4. By gently moving the sliding stage, the fusion products can be removed from the electrode. Move the electrodes out of the droplet. Alignment, adhesion and fusion of egg or central cells with sperm cells are continuously observed under an inverted microscope. Egg–sperm fusions are performed in microdroplets of mannitol (600 or 650 mosmol/kg H_2O). For central cell–sperm fusions, the fusion medium consists of mannitol solution (650 mosmol/kg H_2O). Generally, the osmolality of the fusion medium is lower (50 or 100 mosmol/kg H_2O) than that of the isolation medium. Chemical fusion of central with sperm cells is performed manually using a microneedle for alignment of the two cells in a calcium-containing (5 and 10 mM $CaCl_2$) mannitol solution (600–650 mosmol/kg H_2O), pH 6.0. After the fusion, the fertilized central cells are transferred into the microcapillary by several suction steps of each 15 nL and gently released onto the bottom of the culture dish (*see* **Note 8**).

3.3. Zygote, Embryo and Endosperm Development, Plant Regeneration

1. Place 100 μL ZMS medium in a "Millicell-CM" insert and insert it into a 3.5-cm plastic dish, containing 1.5 mL of a feeder suspension (*see* **Notes 2** and **9**). By using a microcapillary, transfer the fusion products, isolated zygotes and primary endosperm cells into the insert. During the next day, the dish should be placed on a rotary shaker (50–70 rpm) when suspension feeder cells are used. Culture conditions are 26 ± 1.0 °C, a light/dark cycle of 16/8 h, and a light intensity of about 50 μmol/m²/s.
2. Transfer embryos (after about 10–14 days after gamete fusion and growth in liquid culture, minimal size of 0.4 mm length) by use of a Pasteur pipette or a small blade onto 1.5 mL solidified RMS1 regeneration medium in a 3.5-cm plastic dish for a first passage of 2 weeks (*see* **Note 10**). When a coleoptile and roots are formed, transfer the structures for another 1–2 weeks onto 1.5 mL of RMS2 medium. Transfer plantlets into a glass jar containing 50 mL of RMS3 medium. After about another 2 weeks, transfer the maize plants (leaf lengths about 15–20 cm) to soil.
3. Unfertilized and in vitro-fertilized central cells are cultured as described previously for in vitro zygotes but with minor modifications. Central and primary endosperm cells are cultured in inserts (12-mm diameter Millicell-CM dishes) that had been filled with 100 μL of medium. These dishes are located in a 3.5-cm diameter plastic dish filled with 1.5 mL of a maize feeder suspension. For culture, a modified Murashige and Skoog medium is used as described for egg cells. In some experiments, to prevent bursting of floating cells on the surface of the medium, unfertilized and fertilized central cells are transferred onto the membrane of a Millicell-CM dish that had been filled with 100 μL of a semisolidified mannitol solution (600 mosmol/kg H_2O; 0.5 and 0.75%, respectively, ultralow gelling temperature agarose, type IX; Sigma). The cultures are maintained on a rotary shaker at 50 rpm, starting 6 days after cell fusion. The culture conditions

for primary endosperm cells are as described above for zygotes. Six days after the cell fusion, the insert containing the endosperm is transferred into one well of a four-well multidish that had been filled with 300 μL of the previously conditioned medium but without feeder cells. When the endosperms reach a length of approximately 1.5 mm and a width of approximately 0.7 mm (that is, ~11 days after IVF), they are subcultured on 2.0 mL of solidified RMS1 medium in a 3.5-cm diameter plastic dish and maintained under the culture conditions described above.

3.4. Identification of Major Proteins in Egg Cells and Comparison of Their Expression Level Between Egg Cells, Zygotes, Early Embryos and Central Cells

1. Prepare egg cells, zygotes, early embryos and central cells as given in **Subheading 3.1** and wash four times by transferring them into fresh droplets of mannitol solution (650 mosmol/kg H_2O) on cover slips.
2. Transfer 10–50 isolated egg cells, zygotes, early embryos into a 1 μL droplet of SDS sample buffer for SDS–polyacrylamide gel electrophoresis (PAGE) or into a 1μL droplet of lysis buffer for two-dimensional (2D)-PAGE. Transfer one cell central into a 1μL droplet of SDS sample buffer or lysis buffer. Store these samples at −80 °C until use (*see* **Note 11**).
3. Prepare 12.5% SDS-polyacrylamide gels in a small mold (50 × 60 × 1 mm; Atto). For LC-MS/MS analysis, dissolve 75 egg cells in 6 μL of SDS sample buffer applied to the SDS–PAGE. For 2D-PAGE, prepare the isoelectric focusing gel using a thin glass capillary (length, 5 cm; diameter, 1 mm) and dissolve 180 egg cells in 6 μL of lysis buffer applied to the capillary gel. After isoelectric focusing at 100 V for 1 h and subsequently at 300 V for 3 h, incubate gels in SDS–PAGE sample buffer for 10 min with gentle shaking. Separate the proteins in the capillary gel further by 12.5% SDS–PAGE (*see* **Fig. 3**).
4. Incubate the separated SDS–PAGE gels in fixative (50% methanol, 7% acetic acid) with shaking for 3–16 h. Discard fixative and add 50% methanol. After 10-min incubation with shaking, further incubate gels with distilled water (DW) for 10 min. Discard DW and add enhancement solution (0.02% sodium thiosulfate). Incubate for 3 min with shaking, wash gels with DW twice and incubate with freshly prepared silver staining solution (24.7 mg/mL silver nitrate, 0.3% NaOH, 0.11% ammonia) for 30 min. Discard silver staining solution and wash gels with DW twice. Add colour development solution (2% sodium carbonate, 0,04% formaldehyde) and change the solution 5–10 times when the solution turns to brown colour. Stop colour development by exchanging the solution with 7% acetic acid. Gels can be stored at 4 °C in 1% acetic acid for a month. The proteins in the gel used for in-gel tryptic digestion and subsequent analyses with LC-MS/MS are visualized by modified silver staining (*see* **Note 12**).
5. Digest in-gel with trypsin excised protein bands or spots from the SDS–PAGE gel and subject to direct nano-flow LC-MS/MS analysis for protein identification. Perform the chromatography on a nano ESI column (inside diameter, 150 μm × 30

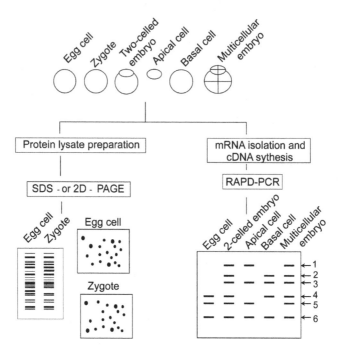

Fig. 3. Identification of proteins (left) and genes (right) expressed in egg, zygotes and/or young embryos. From egg cells, zygotes, two-celled embryos, apical cells, basal cells and early embryos, protein lyzates (left) and mRNAs (right) are prepared. The protein lysates are separated by sodium dodecyl sulphate (SDS–PAGE) or two-dimensional (2D)-PAGE, and silver-stained (left bottom images). cDNAs are synthe-sized from the mRNAs, and the DNA is used as template for RAPD-PCR, followed by separation of the amplified DNA by 1.2% agarose gel electrophoresis. The patterns of DNA bands detected in the gel are then compared among the five cell/embryo types. Based on the detection patterns of the DNA bands showing the same mobility in the gel, the expression profiles can be categorized into six groups. The arrows on the right bottom panel indicate group numbers. Group 1, amplified DNA band (arrow 1) from a putative gene transcript that is up-regulated in the apical cell after fertilization. Group 2, amplified DNA band (arrow 2) from a putative gene transcript that is up-regulated in the basal cell after fertilization. Group 3, amplified DNA band (arrow 3) from a putative gene transcript that is up-regulated in both the apical and basal cells after fertilization. Group 4, amplified DNA band (arrow 4) from a putative gene transcript that is down-regulated only in the apical cell after fertilization. Group 5, amplified DNA band (arrow 5) from a putative gene transcript that is down-regulated only in the basal cell after fertilization. Group 6, amplified DNA band (arrow 6) from a putative gene transcript that is constitutively expressed in all cell/embryo types.

mm) packed with a C18 reversed-phase medium (Mightysil-C18, 3 μm, Kanto Chemicals) using a linear gradient from 0 to 70% acetonitrile in 0.1% formic acid for 35 min at the flow rate of 100 nL/min.

6. Spray the separated peptides directly into a hybrid mass spectrometer equipped with an electrospray source (Q-Tof ultima; Micromass-Waters). Electrospray ionization is carried out at a voltage of 1.5 kV, and MS/MS spectra are automatically acquired in data-dependent mode during the entire run. All MS/MS spectra are correlated by search engine, Mascot program (Matrixscience), against the non-redundant protein sequence database at the National Center for Biotechnology Information (National Institutes of Health) (*see* **Note 13**).

7. For comparison of band patterns between egg cells, zygotes, early embryos and central cells, separate protein lysates from 15 egg cells, 15 zygotes, 15 early embryos or 3 central cells by 12.5% SDS–PAGE (*see* **Note 14**). Incubate the separated SDS–PAGE gels in fixative (30% ethanol and 10% acetic acid) with shaking for 15 min. Discard fixative and incubate the gels with 30% ethanol, 0.2% glutaraldehyde, 500 mM sodium acetate and 13 mM sodium thiosulfate for 10 min with shaking. After washing the gels with DW three times for 3 min, add silver staining solution (6 mM silver nitrate, 0.06% formaldehyde) and incubate for 5 min. Discard silver staining solution, and add colour development solution (250 mM sodium carbonate, 0.03% formaldehyde). Stop the colour development with 50 mM EDTA and 500 mM sodium acetate, pH 7.4 (*see* **Note 12**).

3.5. Identification of Genes that are Up- or Down-Regulated in the Apical or Basal Cell of Two-Celled Embryos and Monitoring Their Expression During Zygote Development by a PCR-Based Approach

1. Isolate egg cells as given in **Subheading 3.1.6., step 6** and prepare and culture in vitro fertilized zygotes as given in **Subheading 3.2., step 3**. Observe cultured zygotes under an inverted microscope at 1 h intervals from 36 to 50 h after fertilization (*see* **Note 15**). When two-celled embryos are observed, transfer them immediately from the culture medium into droplets of mannitol solution (650 mosmol/kg H_2O) containing 1% cellulase Onozuka-RS (Yakult, Tokyo), 0.3% pectolyase Y23 (Kikkoman, Tokyo), 1% hemicellulase (Sigma) and 0.5% driselase (Sigma) on cover slips (*see* **Note 16**).

2. Observe pairs of protoplasts composed of a protoplast from the small apical cell and a protoplast from the large basal cell during treatment with the enzyme solution for 10–20 min (*see* **Note 17**). Gently touch the attachment between the two protoplasts with a very thin glass needle, resulting in the separation of the two protoplasts. These protoplasts are used as isolated apical and basal cells.

3. Wash isolated egg cells, zygotes, apical cells, basal cells, two-celled embryos and multicellular embryos three times by transferring the cells/embryos into fresh droplets of mannitol solution (650 mosmol/kg H_2O) on cover slips.

4. For mRNA isolation, transfer the cells/embryos into 5 μL droplets of the lysis/binding buffer supplied in a Dynabeads mRNA DIRECT™ Micro Kit and stored at −80 °C until use.

5. Prepare mRNA from five cells/embryos using the above-mentioned kit according to the manufacturer's instructions and prepare cDNA from each mRNA using a Super SMART™ PCR cDNA Synthesis Kit according to the manufacturer's instructions (*see* **Note 18**).

6. Use HotStarTaq Mix (25 μL; Qiagen, Hilden) for PCRs with the RAPD primers (1 μM) and 2.5 ng cDNA template from egg cells, apical cells, basal cells, two-celled embryos or multicellular embryos. Conduct PCR under the conditions: 35 cycles of 94 °C for 1 min, 50 °C for 1 min and 72 °C for 2 min.

7. After the electrophoresis of the PCR products by 1.2% agarose gel, compare the profiles of the amplified DNA bands in gels among the five cell/embryo types. Based on the detection patterns of the DNA bands showing the same mobility in the gels, the expression patterns can be categorized into six groups as shown on the right panel in **Fig. 3**.

8. Subclone the amplified DNA bands into the pGEM-T Easy vector (Promega) and sequence. Design specific primers for selected clones and conduct PCR with the specific primers to verify the expression profile in cells/embryos (*see* **Note 19**).

3.6. Immunocytochemical Techniques Adapted to Single-Egg Cells and Zygotes

1. Transfer isolated egg cells or cultured zygotes into "Millicell-CM" inserts containing 100 μL mannitol solution (650 mosmol/kg H_2O) in 3.5-cm diameter plastic dishes. The transfer of individual cells is performed using a microcapillary that is connected to a micropump via a hydraulic system (*see* **Fig. 4**). In order to continuously monitor the cells, all procedures in this protocol are performed under an inverted microscope.

2. Reduce the amount of mannitol solution in the Millicell-CM inserts to prevent dilution of the fixative during the subsequent step of the procedure.

3. Subsequently, transfer Millicell-CM inserts containing the cells into other 3.5-cm diameter plastic dishes that have previously been filled with 4 mL freshly prepared fixative solution.

4. Remove remaining mannitol solution in the Millicell-CM inserts and add the fixative into the inserts by using a microcapillary. The semi-permeable membrane of the Millicell-CM inserts allows the exchange of the fixative between the two compartments.

5. After fixation for 20 min, transfer cells into another new Millicell-CM insert containing 100 μL MSB. This new insert has previously been placed in a 3.5-cm diameter plastic dish filled with 4 mL MSB. These cells can be kept in a refrigerator until the subsequent detection steps.

6. Treat glass cover slips (24 × 40 mm) with poly-L-lysine to facilitate cellular adhesion to the surface of the cover slips. Using adhesive tapes, reservoirs are built on the cover slips in order to keep the solutions in place (*see* **Fig. 4**).

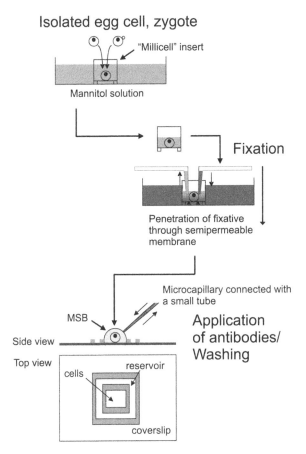

Fig. 4. Cytochemistry, fixation and antibody application adapted to single egg cells and zygotes.

7. Mount fixed cells in a 50 μL MSB droplet on the cover slips by using a microcapillary. The cells move downward into the droplet due to the effect of gravity and adhere to the cover slip.

8. Incubate cells for 1 h in a 50-μL droplet that contains the anti-α-tubulin monoclonal antibody diluted 1:100 in MSB supplemented with 0.25% (w/v) BSA. Place cover slips on moist papers in a Petri dish (12 cm diameter) during the subsequent successive incubation periods.

9. Subsequently, the cells are washed with MSB by using a microcapillary.

10. After three washes with MSB, incubate cells for 1 h with the FITC-conjugated anti-mouse IgG. This antibody is diluted 1:100 in MSB supplemented with 0.25% (w/v) BSA and incubated at room temperature.

11. Subsequently, wash cells again three times with MSB.

12. Treat cells with DAPI for 15 min to stain the DNA.

13. Following three washes with MSB, deliver an antifading solution in MSB on the cover slips.
14. When zygotes are used after more than 20 h of culturing, transfer them onto a cover slip and treat with an enzyme solution for cell wall digestion for 10 min after fixation. After three washes with MSB, the antibodies are applied as described above.
15. Examine cells under an epifluorescence inverted microscope.
16. Perform immunological control experiments by omitting the primary antibody in one experiment, by omitting the secondary antibody and by omitting both.

4. Notes

1. The procedure is line independent. By using well-growing and well-watered donor plants, in maize, routinely 20–40 egg cells can be isolated per person per day, and the same number of fusion products can be obtained. Under optimal conditions, up to 60 zygotes can be created by one person per day. Three to eight central cells can be isolated from approximately 160 tissue pieces within 2–3 h and fused with sperm cells. Compared with egg cell isolation, some modifications are required for the efficient isolation of maize central cells. Whereas the treatment of nucellar tissue with cell wall degrading enzymes for efficient egg cell isolation is not >30 min, a longer duration for this treatment (45 min) is optimal for central cell isolation. In addition, it is advantageous to use a cell wall degrading enzyme mixture with an osmolarity higher than that used for egg cell isolation. Plasmolysis results in the separation of the central cell membrane from the embryo sac wall, which is favorable to improve the manual isolation step. After this treatment, central cells are manually isolated directly in the incubation dish with glass needles under an inverted microscope. Some nucellus cells are removed only from the micropylar end near the embryo sac. Beginning at the chalazal end, near the antipodal cells, the central cell is pushed toward the micropylar end with a glass needle. Here, the cell is liberated and becomes spherical. Occasionally, it is possible to isolate a unit of an egg, two synergids and a central cell. Subsequently, the adherent egg and synergids can be separated manually from the central cell by using a microneedle.
2. Feeder suspension cells need not to be embryogenic, but for a positive feeding effect, they should grow exponentially. It can be useful to subculture suspension cells twice a week. Microspore cultures also can be used as a feeder system *(4)*. Although the feeding effect can be also achieved with nurse cells of a different species than the zygote, make sure that culture conditions meet the requirements of both the zygotes and the feeder cells. When suspension cells are used for feeding, these have to be adapted (usually several passages before use) to a higher osmolarity than normally used for these cells, to meet the requirement for protoplasts.
3. The use of mannitol solution (650 mosmol/kg H_2O after autoclaving) in this step can be advantageous in improving cell fusion.

4. The concentration of cell wall degrading enzymes should be kept low and treatment should be short to avoid spontaneous fusion of the cells of the embryo sac when ovule tissue is softened with such enzymes.

5. Isolated sperm cells are short-lived. Therefore, it is important to provide conditions that guarantee the maintenance of good quality of living sperm cells during the time necessary to perform gamete fusion. For example, whereas sperm cells of maize can be used for fusion within 30–45 min after isolation *(9,10)*, sperm cells of wheat are often useful for fusion for only few minutes after isolation *(8,11)*. Occasionally, the two sperm cells spontaneously fuse after release from a pollen grain. To avoid the use of such a fusion product, select sperm cells only when two are visible in close proximity to each other and separated from others. If available, use oval or spindle-shaped sperm cells for fusion because these fuse more efficiently with egg and central cells than round ones *(35)*.

6. The turgor of the cells should be high enough for efficient cell fusion. Avoid shifts from low to higher osmolarities between isolation and fusion media. Use the same medium preparation for isolation and fusion. Microdroplets of media always should be covered by oil, otherwise the osmolarity of the medium will fast increase with deleterious effects to the cells.

7. In case no cell fusion occurs, lower the distance of the two electrodes and pulse again. Occasionally, some remnant cell wall material is attached to the egg membrane which prevents fusion. This can be overcome by rotation of the egg cell to establish a new fusion site. Another reason for non-fusion can be that the turgor of the protoplasts is too low. If this is the case, transfer the protoplasts into a fusion medium with a slightly lower osmolarity. When protoplasts reiteratively burst after the applied pulse, the osmolarity of the fusion medium is too low or the electrodes are not well adjusted. Use dielectric alignment conditions of 1 MHz, 30 V/cm and DC pulses of 0.3 kV/cm, 50 µs for the fusion combination of cells of the same size, for example, egg + egg or egg + somatic protoplast.

8. There are three reports describing calcium-mediated egg-sperm fusion *(13–15)*. Also in this procedure, the turgor of the protoplasts is important. Thus, determine the optimum osmolarity of the fusion medium. The alignment of the two gametes can be performed by a glass needle in a microdroplet. When adhesion has occurred, egg and sperm cells or central cells and sperm cells can be fused in mannitol solution (about 400–500 mosmol/kg H_2O) containing 1–50 mM calcium ($CaCl_2$). This time-consuming method has to be improved for efficient routine work.

9. Isolated gametes and zygotes can be encapsuled in droplets of agarose-containing mannitol solution (ultra-low gelling temperature agarose, type IX, Sigma, 1%). After cell transfer, the droplets are solidified for 10 min at 6 °C. The droplets can be fixed on small plastic pieces, which can be useful for example for fixation, electron microscopical studies and transportation of single cells to another laboratory for further analyses *(12,18)*. To follow development at a fixed

position, zygotes and primary endosperm cells can be embedded in agarose-containing medium, previously filled into the "Millicell-CM" inserts (100 μL, ultra-low gelling temperature agarose, type IX, Sigma, 1%).

10. Transfer embryos from liquid culture medium (ZMS) onto solid RMS1 medium supplemented with 2,4-D (2 mg/L) for two to three passages to obtain embryos of later stages.

11. Maximum volume of loading sample for one-dimensional-PAGE (SDS–PAGE) and 2D-PAGE (isoelectric focusing) are 8 and 10 μL, respectively.

12. When the SDS–PAGE gel (50 × 60 × 1 mm) is stained by a conventional silver staining protocol, protein bands can be detected using 15 egg cells. However, approximately five times more cells are needed to detect protein bands in SDS–PAGE gels, when silver staining methods modified for LC-MS/MS analysis is employed.

13. Although the LC-MS/MS system has extremely high sensitivity, only protein bands with strong intensity can be analyzed due to limitation of sample amounts. This analytical limitation confirms that the identified proteins are not derived from minor proteins overlapping with the major proteins in the gel, but from major proteins themselves.

14. By comparing the intensity of protein bands from egg cells with that of the molecular weight marker co-migrated on the gel, the amount of protein in an egg cell is roughly estimated to be 100–200 picogram. The protein amount in a central cell is approximately equivalent with that from five egg cells.

15. In our IVF/culture system, zygotes generally divide into two-celled embryos at 40–50 h after fertilization.

16. Driselase is very effective in separation of apical and basal cells from the two-celled embryo.

17. Long incubation of the embryo with the enzymatic solution often results in fusion of a pair of protoplasts from the apical and basal cells.

18. The numbers of cells for mRNA isolation and subsequent cDNA synthesis are usually 5–20 cells depending on efficiency of isolation/preparation of cells. Because of difficulty of preparation of the apical and basal cells, we used minimum number of cells as starting materials.

19. Using 180 kinds of RAPD primers, 17, 35, 14, 35 and 1 bands were detected as the gene products in group 1, 2, 3, 4 and 5, respectively (*see* **Fig. 3**).

References

1. Kranz E, Lörz H. In vitro fertilization with isolated, single gametes results in zygotic embryogenesis and fertile maize plants. Plant Cell 1993;5:739–746.
2. Kranz E, von Wiegen P, Quader H, Lörz H. Endosperm development after fusion of isolated, single maize sperm and central cells in vitro. Plant Cell 1998;10:511–524.
3. Goldberg RB, de Paiva G, Yadegari R. Plant embryogenesis: zygote to seed. Science 1994;266:605–614.

4. Holm PB, Knudsen S, Mouritzen P, Negri D, Olsen FL, Roué C. Regeneration of fertile barley plants from mechanically isolated protoplasts of the fertilized egg cell. Plant Cell 1994;6:531–543.

5. Kovács M, Barnabás B, Kranz E. The isolation of viable egg cells of wheat (*Triticum aestivum* L.). Sex Plant Reprod 1994;7:311–312.

6. Katoh N, Lörz H, Kranz E. Isolation of viable egg cells of rape (*Brassica napus* L.). Zygote 1997;5:31–33.

7. Kumlehn J, Brettschneider R, Lörz H, Kranz E. Zygote implantation to cultured ovules leads to direct embryogenesis and plant regeneration of wheat. Plant J 1997;12:1473–1479.

8. Kranz E, von Wiegen P, Lörz H. Early cytological events after induction of cell division in egg cells and zygote development following *in vitro* fertilization with angiosperm gametes. Plant J 1995;8:9–23.

9. Kranz E, Bautor J, Lörz H. *In vitro* fertilization of single, isolated gametes of maize mediated by electrofusion. Sex Plant Reprod 1991;4:12–16.

10. Kranz E, Bautor J, Lörz H. Electrofusion-mediated transmission of cytoplasmic organelles through the *in vitro* fertilization process, fusion of sperm cells with synergids and central cells, and cell reconstitution in maize. Sex Plant Reprod 1991;4:17–21.

11. Kovács M, Barnabás B, Kranz E. Electro-fused isolated wheat (*Triticum aestivum* L.) gametes develop into multicellular structures. Plant Cell Rep 1995;15: 178–180.

12. Faure J-E, Mogensen HL, Dumas C, Lörz H, Kranz E. Karyogamy after electro-fusion of single egg and sperm cell protoplasts from maize: cytological evidence and time course. Plant Cell 1993;5:747–755.

13. Kranz E, Lörz H. *In vitro* fertilisation of maize by single egg and sperm cell protoplast fusion mediated by high calcium and high pH. Zygote 1994;2:125–128.

14. Faure J-E, Digonnet C, Dumas C. An *in vitro* system for adhesion and fusion of maize gametes. Science 1994;263:1598–1600.

15. Khalequzzaman M, Haq N. Isolation and in vitro fusion of egg and sperm cells in *Oryza sativa*. Plant Physiol Biochem 2005;43:69–75.

16. Sun M, Yang H, Zhou C, Koop H-U. Single-pair fusion of various combinations between female gametoplasts and other protoplasts in *Nicotiana tabacum*. Acta Bot Sin 1995;37:1–6.

17. Antoine A-F, Faure J-E, Dumas C. Feijó JA. Differential contribution of cytoplasmic Ca^{2+} and Ca^{2+} influx to gamete fusion and egg activation in maize. Nat Cell Biol 2001;3:1120–1123.

18. Tirlapur UK, Kranz E, Cresti M. Characterization of isolated egg cells, *in vitro* fusion products and zygotes of *Zea mays* L. using the technique of image analysis and confocal laser scanning microscopy. Zygote 1995;3:57–64.

19. Scholten S, Lörz H, Kranz E. Paternal mRNA and protein synthesis coincides with male chromatin decondensation in maize zygotes. Plant J 2002;32: 221–231.

20. Koop H-U, Schweiger, H-G. Regeneration of plants after electrofusion of selected pairs of protoplasts. Eur J Cell Biol 1985;39:46–49.

21. Spangenberg G, Koop H-U. Low density cultures: microdroplets and single cell nurse cultures. In: Lindsey K, ed. Plant Tissue Culture Manual, A10. Dordrecht: Kluwer Academic Publishers, 1992:1–28.

22. Hoshino Y, Scholten S, von Wiegen P, Lörz H, Kranz E. Fertilization-induced changes in the microtubular architecture in the maize egg cell and zygote – an immunocytochemical approach adapted to single cells. Sex Plant Reprod 2004;17:89–95.

23. Okamoto T, Scholten S, Lörz H, Kranz E. Identification of genes that are up- or down-regulated in the apical or basal cell of maize two-celled embryos and monitoring their expression during zygote development by a cell manipulation- and PCR-based approach. Plant Cell Physiol 2005;46:332–338.

24. Okamoto T, Kranz E. In vitro fertilization- a tool to dissect cell specification from a zygote. Curr Sci 2005;89:1861–1869.

25. Okamoto T, Higuchi K, Shinkawa T, Isobe T, Lörz H, Koshiba T, Kranz E. Identification of major poteins in maize egg cells. Plant Cell Physiol 2004;45: 1406–1412.

26. Okamoto T, Kranz E. Major proteins in plant and animal eggs. Acta Biologica Cracoviensia, Series Botanica, 2005;47:17–22.

27. Richert J, Kranz E, Lörz H, Dresselhaus T. A reverse transcriptase-polymerase chain reaction assay for gene expression studies at the single cell level. Plant Sci 1996;114:93–99.

28. Dresselhaus T, Lörz H, Kranz E. Representative cDNA libraries from few plant cells. Plant J 1994;5:605–610.

29. Dresselhaus T, Hagel C, Lörz H, Kranz E. Isolation of a full-length cDNA encoding calreticulin from a PCR library of *in vitro* zygotes of maize. Plant Mol Biol 1996;31:23–34.

30. Lê Q, Gutièrrez-Marcos J, Costa L, Meyer S, Dickinson H, Lörz H, Kranz E, Scholten S. Construction and screening of subtracted cDNA libraries from limited populations of plant cells: a comparative analysis of gene expression between maize egg cells and central cells. Plant J 2005;44:167-178.

31. Kranz E, Dresselhaus T. *In vitro* fertilization with isolated higher plant gametes. Trends Plant Sci 1996;1:82–89.

32. Rougier M, Antoine AF, Aldon D, Dumas C. New lights in early steps of *in vitro* fertilization in plants. Sex Plant Reprod 1996;9:324–329.

33. Kranz E, Kumlehn J. Angiosperm fertilization, embryo and endosperm development in vitro. Plant Sci 1999;142:183–197.

34. Kranz E. *In vitro* fertilization of maize mediated by electrofusion of single gametes. In: Lindsey K, ed. Plant Tissue Culture Manual, E1. Dordrecht: Kluwer Academic Publishers, 1992:1–12.

35. Kranz E. *In vitro* fertilization with isolated single gametes. In: Hall R, ed. Methods in Molecular Biology: Plant Cell Culture Protocols, vol. 111. Totowa, NJ: Humana Press Inc., 1999:259–267.

36. Murashige T, Skoog F. A revised medium for rapid growth and bioassays with tobacco tissue cultures. Physiol Plant 1962;15:473–497.
37. Brown RC, Lemmon BE. Methods in plant immunolight microscopy. Methods Cell Biol 1995;49:85–107.
38. Schliwa M, van Blerkom J. Structural interaction of cytoskeletal components. J Cell Biol 1981;90:222–235.

5

In Vitro Culture of *Arabidopsis* Embryos

Michael Sauer and Jiří Friml

Summary

Embryogenesis of *Arabidopsis thaliana* follows a nearly invariant cell division pattern and provides an ideal system for studies of early plant development. However, experimental manipulation with embryogenesis is difficult, as the embryo develops deeply inside maternal tissues. Here, we present a method to culture zygotic *Arabidopsis* embryos in vitro. It enables culturing for prolonged periods of time from the first developmental stages on. The technique omits excision of the embryo by culturing the entire ovule, which facilitates the manual procedure. It allows pharmacological manipulation of embryo development and does not interfere with standard techniques for localizing gene expression and protein localization in the cultivated embryos.

Key Words: Embryogenesis; in vitro culture; pharmacological manipulation; phytohormone treatment.

1. Introduction

The process of embryogenesis is of fundamental interest to developmental biology, as the main axes of symmetry are created and self-perpetuating patterns of cell identities are laid out in its course. To study certain aspects of embryogenesis, it can be necessary to culture embryos outside the maternal tissue, creating a more controlled in vitro environment. Abiotic parameters such as light or temperature can be regulated, or, more specifically, administration of specific inhibitors, phytohormones, or even the induction of chemically inducible transgenic expression systems is possible in in vitro culture (IVC). IVC of excised *Arabidopsis* embryos has so far been demonstrated only in embryos after the heart stage, when the fundamental patterning decisions already have been

From: *Methods in Molecular Biology, vol. 427: Plant Embryogenesis*
Edited by: M. F. Suárez and P. V. Bozhkov © Humana Press, Totowa, NJ

taken *(1,2)*. Reports on somatic embryogenesis are more widespread *(3,4)*, but this method suffers from that it is unclear to which extent somatic and zygotic embryogenesis are equivalent and also lacks well-defined early stages.

The culture of embryos inside their ovules is a widespread method in the plant field, although until recently, there were no studies with a focus on development in *Arabidopsis*. There are protocols to rescue specific hybrids or mutants, for example, for grapes *(5)*, for wheat *(6)*, or applications where the development of the ovule itself is analyzed, for example, cotton fibre development *(7)*. Recently, we reported an ovule-based culture system for the study of *Arabidopsis* embryogenesis *(8)*, which is also suitable for the culture of very early embryos. Developmental aberrations induced by this culture method are acceptably low. We further showed that administration of physiologically active drugs or hormones is possible and that the method is compatible with various subsequent analytical methods, such as fluorescence microscopy *(9,10)* or immunocytochemistry *(8)*. In this chapter, we provide a slightly improved protocol, along with basic recommendations for microscopic embryo analysis.

2. Materials

2.1. In Vitro Culture

1. IVC medium: 5% (w/v) sucrose, 2.2 g/L Murashige and Skoog salts in water (*see* **Note 1**), adjust pH with KOH to 5.9.
2. Only for long-term culture (>5 days): Arabidopsis Medium (AM) – 1% (w/v) sucrose, 2.2 g/L Murashige and Skoog salts, 0.5 g/L MES (2-(N-Morpholino)ethanesulfonic Acid) buffer in water, adjust pH with KOH to 5.9.
3. Gelling agent: For IVC, Phytagel (Sigma); For AM, research grade agar (e.g., from Serva).
4. Glutamine stock solution: glutamine is dissolved in water at 16 g/L, this solution should be prepared fresh each time.
5. Fine syringe needles (0.3 mm diameter, e.g., insulin syringes), double-sided adhesive tape, dissecting scope.

2.2. Microscopic Analysis

1. Clearing solution: 66% (w/v) chloralhydrate, 24% (v/v) water, 10% (w/v) glycerol.
2. Glycerol solution: 5% (v/v) glycerol in water.

3. Methods

3.1. Preparation of the Culture Plates

1. Fill IVC in flasks for autoclaving and add 0.3% (w/v) Phytagel (*see* **Note 2**). Autoclave 10 min (*see* **Note 3**), allow to cool to about 50 °C, then sterile filter 25

mL/L glutamine stock solution into the medium. If desired, add drugs, inhibitors, hormones, and so on now. Pour the medium into 50-mm diameter Petri dishes and allow to solidify. Plates can be stored for several weeks at 4 °C.

2. Only for long-term culture (>5 days): Fill AM in flasks for autoclaving and add 0.8% (w/v) agar. Autoclave, (if desired, let cool and add drugs, inhibitors, hormones, etc.). Pour medium into 50-mm diameter Petri dishes and allow to solidify.

3.2. Start of In Vitro Culture

1. Surface sterilize a dissecting microscope with 70% EtOH, prepare microscope slides with strips of double-sided adhesive tape and sterilize likewise (The tape will become sticky again when dry) (*see* **Note 4**).
2. From *Arabidopsis* plants, take siliques of the appropriate stage and surface sterilize them by a 15-s dip in 70% EtOH. Let them dry on tissue paper (*see* **Note 5**) and stick them to the slides with double adhesive tape.
3. Under a dissecting scope, cut open the silique along the replum with syringe needles, stick the sides of the valves onto the tape and carefully scrape out the ovules with sterile forceps, taking care not to damage the ovules.
4. Immediately transfer the ovules to the surface of a culture plate. Do not submerge the ovules in the medium. After transfer of the ovules of three to four siliques, close the culture plate and seal with laboratory film (*see* **Note 6**).
5. Place the culture plates upside down (ovules hanging from the medium) in a plant growth chamber. Shield from excessive light by one or two sheets of plain white paper.
6. Culture the ovules at desired temperature according to your experimental setup (*see* **Note 7**). If you need to culture for more than 5 days, follow steps given in **Subheading 3.3.**

3.3. Transfer to AM Medium for Long-Term Culture

1. After 5 days of culture, open the plates under sterile conditions. Preferentially using a dissecting scope, transfer the ovules from IVC plates to AM plates, again, taking care not to damage or to submerge them in the medium. At this stage, you may want to exclude obviously damaged or aborted ovules (*see* **Note 8**).
2. Seal the AM culture plates with laboratory film and continue culturing them as described in **Subheading 3.2., step 5** and following.

3.4. Microscopic Morphological Embryo Analysis

1. Open the culture plates. Under a dissecting scope, transfer the ovules to microscope slides into a drop of clearing solution.
2. Cover with a standard cover slip, then incubate the slide at 37 °C for about 20 min (*see* **Note 9**).

3. Analyze the slides preferentially under a microscope with differential interference contrast (DIC) optics (sometimes referred to as 'Nomarski' optics).

3.5. Microscopic Analysis of Embryos Expressing Fluorescent Proteins

1. Open the culture plates. Under a dissecting scope, transfer the ovules from the medium to microscope slides into a drop of glycerol solution (*see* **Note 10**).
2. Under a dissecting scope, preferentially with transmitted light, use syringe needles to cut open or tear the ovules. Cover with a standard cover slip, then apply gentle pressure (e.g., with forceps or lead pencil) on the ovules in order to release the embryos (*see* **Note 11**).
3. Analyze the slides with a fluorescence microscope (epifluorescence widefield microscope, confocal laser scanning microscope, etc.). Use the appropriate excitation and emission filter settings for the fluorescent protein of your study.

3.6. Embryo Analysis with Histochemical Methods

Histochemical methods, such as detection of β-Glucuronidase activity (GUS staining) or immunocytochemistry, are possible without restrictions. The embryos inside the cultured ovules can be treated just like freshly isolated embryos. The ovules should only be washed two times for 5 min in water, to remove any traces of culture medium, prior to any step of the histochemical method.

4. Notes

1. Throughout this chapter, 'water' refers to bidistilled or reverse osmosis water.
2. If preparing more than one flask of medium, Phytagel, as any gelling agent, should be added directly to the flasks before autoclaving, as it tends to sediment quickly.
3. Do not autoclave longer, as Phytagel based media may not solidify any more if 'overcooked'. Also, do not remelt this medium. AM medium with agar as a gelling agent is less sensitive to this.
4. To reduce the risk of contamination, you can shield the scope from airborne contaminants with a simple cardboard box, but this is not absolutely required. Of course, if a laminar flow hood is available, it is preferable to set up the scope there.
5. Do not take too many siliques at once (three to six at once are fine), as they will dry out rapidly. Dried out siliques decrease embryo survival and pose a greater risk of ovule damage.
6. It is often a good idea to set aside a small number of ovules in order to determine the developmental stage at culture start. Simply transfer a couple of ovules on a slide with a small drop of clearing solution (see also **Subheading 3.4.**)

7. Culture temperature and duration very much depend on the actual experimental setup. As a guideline, under conditions of 23 °C and dim light, embryogenesis proceeds from four-cell stage to maturity (bent cotyledon stage) in 7 days. Lower temperatures slow down the process.

8. Ovules, which have been damaged in the handling process, will after several days turn brown or be flattened instead of round and pale white (or greenish, for older embryos). These ovules should not be taken into consideration in the final analysis as ovule abortion in culture early on will lead to aberrant embryo development. For ovules, which have been cultured only for short durations (<3 days), it may be difficult to discern damaged ovules from healthy ones. In this case, larger sample numbers and a thorough statistic analysis are advisable.

9. Incubating in clearing solution at 37 °C helps in the clearing process. If results are unsatisfactorily, prolong this step, or consider isolation of the embryos from the ovules, as described in **Subheading 3.5., step 2**, only with clearing solution instead of glycerol.

10. Clearing solution is incompatible with many fluorescent probes. For example, green fluorescent protein fluorescence is quenched within seconds.

11. The reasons for isolating embryos from their ovules are that the ovules are not translucent and, above that, have high amounts of autofluorescence in nearly all parts of the visible spectrum, which makes fluorescent observation of embryos inside their ovules virtually impossible.

Acknowledgments

This work was funded by the Volkswagen Stiftung.

References

1. Blilou I, Xu J, Wildwater M, Willemsen V, Paponov I, Friml J, Heidstra R, Aida M, Palme K, Scheres, B. The PIN auxin efflux facilitator network controls growth and patterning in *Arabidopsis* roots. Nature 2005;433:39–44.

2. Harding EW, Tang W, Nichols KW, Fernandez DE, Perry SE. Expression and maintenance of embryogenic potential is enhanced through constitutive expression of AGAMOUS-Like 15. Plant Physiol 2003;133:653–663.

3. Mordhorst A, Voerman K, Hartog M, Meijer E, van Went J, Koornneef M, de Vries S. Somatic embryogenesis in *Arabidopsis* thaliana is facilitated by mutations in genes repressing meristematic cell divisions. Genetics 1998;149:549–563.

4. Luo Y, Koop H. Somatic embryogenesis in cultured immature zygotic embryos and leaf protoplasts of *Arabidopsis* thaliana. Planta 1997;202:387–396.

5. Cain D, Emershad R, Tarailo R. In-ovulo embryo culture and seedling development of seeded and seedless grape (Vitis vinefera L). Vitis 1983;22:9–14.

6. Kumlehn J, Schieder O, Lörz H. In vitro development of wheat (Triticum aestivum L.) from zygote to plant via ovule culture. Plant Cell Reports 1997;16:663–667.

7. Beasley CA. In vitro culture of fertilized cotton ovules. Bioscience 1971;21: 906–907.

8. Sauer M, Friml J. In vitro culture of *Arabidopsis* embryos within their ovules. Plant J 2004;40:835–843.

9. Friml J, Vieten A, Sauer M, Weijers D, Schwarz H, Hamann T, Offringa R, Jurgens G. Efflux-dependent auxin gradients establish the apical-basal axis of Arabidopsis. Nature 2003;426:147–153.

10. Weijers D, Sauer M, Meurette O, Friml J, Ljung K, Sandberg G, Hooykaas P, Offringa R. Maintenance of embryonic auxin distribution for apical-basal patterning by PIN-FORMED-dependent auxin transport in Arabidopsis. Plant Cell 2005;17:2517–2526.

6

Culture and Time-Lapse Tracking of Barley Microspore-Derived Embryos

Simone de F. Maraschin, Sandra van Bergen, Marco Vennik, and Mei Wang

Summary

Barley microspore embryogenesis represents an attractive system to study stress-induced cell differentiation and is a valuable tool for efficient plant breeding. In contrast to zygotic embryogenesis, all developmental stages are freely accessible at a large scale for observation, molecular analysis and manipulation techniques. In barley, there is a high percentage of microspores that become embryogenic after stress treatment in a mannitol solution. These microspores have the capacity to follow an embryogenic route in both liquid and solid cultures, yielding up to 10% of embryos. In this protocol, we describe three different culture systems for obtaining barley microspore-derived embryos, where embryos develop in liquid medium, on top of a solid medium layer or immobilized in a thin layer of agarose. While liquid culture systems allow the generation of large amounts of embryos for molecular analysis, solid culture systems are the ultimate tool for probing embryo development.

Key Words: Barley; androgenesis; microspore embryogenesis; cell tracking; liquid and solid culture.

1. Introduction

Barley (*Hordeum vulgare* L.) microspores can be diverted from their normal gametophytic development towards a sporophytic route, leading to the formation of haploid embryos and ultimately doubled-haploid plants. This process is called androgenesis or microspore embryogenesis and is widely used to generate homozygous lines for breeding purposes. On top of its application in breeding, barley androgenesis is an attractive system to study stress-induced cell

From: *Methods in Molecular Biology, vol. 427: Plant Embryogenesis*
Edited by: M. F. Suárez and P. V. Bozhkov © Humana Press, Totowa, NJ

differentiation and plant embryogenesis due to the high regeneration efficiencies that are obtained when isolated microspores are cultured *(1)*.

For switching on the genetic program of microspores from the gametophytic towards the sporophytic pathway, a trigger is necessary. In barley, androgenesis can be efficiently induced at the mid-late to late (ML-L) uninucleate stage of microspore development either via a cold pre-treatment of tillers or an incubation of anthers in 0.37 M mannitol solution *(2,3)*. The latter, which is a combination of osmotic and starvation stress, has been proven to be more successful in our laboratory than a cold pre-treatment of tillers. In addition, microspores that have been exposed to mannitol treatment have shown to be amenable to both liquid and solid culture. Liquid culture systems allow the production of large amounts of embryos which have proven to be excellent material for morphological, molecular and biochemical studies. Based on this system, several markers have been found to be associated with the embryogenic competence of microspores after stress treatment, such as the enlarged morphology of embryogenic microspores *(2)* and the expression pattern of specific genes and proteins *(4–6)*.

Nevertheless, the asynchronous development of microspore-derived embryos in liquid culture represents a challenge for the establishment of developmental patterns. Evidence for developmental patterns can only be obtained by pursuing individuals throughout their development from an individual microspore to embryo formation. This can be achieved by culturing microspores on top of a solid medium layer or embedded in solid medium. The latter easily allows the establishment of a monitoring system which is suitable for cell tracking. Cell tracking systems have enabled the observation of key developmental aspects during the embryogenic process of barley microspore-derived embryos, such as the formation of star-like morphology. Star-like microspores represent a transitory stage between enlarged microspores after stress and the initiation of cell divisions *(7–9)*. Furthermore, by combining cell fate markers with cell tracking, the precise site for exine wall rupture has been identified. It is situated at the opposite side of the pollen germ pore, and its rupture involves the elimination of a generative cell domain by programmed cell death *(10)*.

In this chapter, we present the procedures used for culturing embryogenic barley microspores from the winter two-rowed cultivar Igri in both liquid and solid media. In the liquid system presented here, an average of 7–10% of the enlarged microspores develops into embryos after 21 days of culture. These microspore-derived embryos are fully capable of regenerating fertile, double-haploid plants *(2,4)*. This system is relatively easy and allows the development of large amounts of embryos which can be used for various purposes. The two solid systems presented involve embryo development in a thin layer of agarose gel or on top of a solid medium layer. Although both are comparable to the liquid system regarding their efficiencies, embryo development is relatively slower in solid cultures *(7)*. While the one that involves the immobilization of microspores in a thin layer of agarose is most suitable for cell tracking, the growth of microspore-derived embryos on top of a solid medium layer opens up new opportunities for manipulation techniques.

2. Materials

2.1. Equipment

Except stated otherwise, all the procedures described for the pre-treatment and culture of isolated barley microspores must be carried out under sterile conditions and with sterilized material.

1. Sterile Petri dishes 145 × 20 mm, 60 × 15 mm, 35 × 10 mm (Ø× height).
2. Sterile forceps and fine-tipped forceps.
3. Sterile glass beakers (100 mL, 250 mL).
4. Binocular stereoscope, upright light microscope, inverted light and UV microscope coupled to a digital imaging and camera system.
5. Glass slides and cover slips.
6. Parafilm.
7. Waring blender and autoclaved MC2 stainless cups (37–110 mL).
8. Autoclaved Teflon rod.
9. Sterile 3-mL plastic pipets, glass Pasteur pipets, micro-pipets and tips, disposable 1- to 5-mL plastic pipets, Pipetboy pippetor.
10. Sterile 10-mL centrifuge tubes, sterile 1-mL Eppendorf tubes.
11. Autoclaved 110-µm nylon mesh coupled to a metal cylinder fitting the top of 100- to 250-mL beaker cup.
12. Fuchs–Rosenthal counting chamber.
13. Autoclaved 500-µm nylon meshes pre-cut to fit into 8-well Lab-Tek II (Nalgene Nunc International).
14. Eight-well Lab-Tek II chambers (Nalgene Nunc International).
15. One-centimetre round sterile filter papers.

2.2. Culture Media and Stock Solutions

1. Ethanol (v/v), 70%.
2. Sucrose solution (w/v), 15%: autoclave it at 121 °C for 20 min and store it at 4 °C.
3. Fluorescein diacetate (FDA) stock solution, 0.04 mg/mL, in 100% ethanol: sterilize it by filtration and store it at –20 °C.
4. Mannitol, 0.37 M, in CPW salt solution: this solution is used for barley anther pre-treatment (*see* **Table 1**). The pH and the osmolality of the mannitol solution should be around 5 and 0.440 mOsm/kg, respectively (*see* **Note 1**). The mannitol solution is autoclaved at 121 °C for 20 min and can be stored at room temperature.

Table 1
Composition of 0.37 M Mannitol in CPW Salt Solution

Component	mg/L
$CaCl_2 \cdot 2H_2O$	1480
KNO_3	101
$MgSO_4 \cdot 7H_2O$	246
KH_2PO_4	27.2
$CuSO_4 \cdot 5H_2O$	0.025
KI	0.1575
Mannitol	67,400

5. Medium I: this is the medium used for the culture of mannitol-treated isolated barley microspores in liquid and solid systems, as well as for cell tracking (*see* **Table 2**). For liquid culture, the osmolality of the solution should be 350 mOsm/kg (*see* **Note 2**). The pH is set to 5.6 with 0.1 KOH and the solution is sterilized by filtration. This medium can be stored at –20 °C (*see* **Note 3**). For cell tracking, prepare a double-concentrated stock solution of medium I, sterilize it by filtration and store it at –20 °C. For solid culture, a final concentration of 0.6% agarose (Sea Kem LE) and 10^{-6} M of 2,4-D are added to medium I. This is done by preparing a double-concentrated stock solution of medium I and 2,4-D, sterilizing it by filtration and then mixing it with an equal volume of autoclaved double-concentrated agarose solution.

6. Medium II: this medium is used for the germination of microspore-derived embryos into plantlets (*see* **Table 2**). The pH of the medium is set to 5.8 and the medium is autoclaved at 121 °C for 20 min for sterilization.

7. Two percent low-melting type Sea Plaque agarose stock solution: Autoclave it at 121 °C for 20 min and store it at room temperature.

3. Methods

3.1. Growth of Donor Plants

Donor plants of winter two-rowed barley (*Hordeum vulgare* L. cv Igri, Landbouw Bureau Wiersum) are grown in a growth chamber at 12 °C with a 16-h light regime (24,000 lux) and 80% relative humidity.

3.2. Anther Pre-Treatment

1. Tillers with interligule length from 4 to 6 cm are excised just beneath the inflorescence node (*see* **Note 4** and **Fig. 1A**). Tillers are kept in a beaker containing demi-water at 4 °C until they are further processed (*see* **Note 5**).

Table 2
Composition of Medium I and II

Component	Medium I mg/L	Medium II mg/L
Macroelements		
KNO_3	1750	1900
NH_4NO_3	165	–
KH_2PO_4	200	170
$MgSO_4 \cdot 7H_2O$	350	180.54
$CaCl_2$	450	332.02
Microelements		
FeNaEDTA	40	36.70
$MnSO_4 \cdot H_2O$	15	16.9
H_3BO_3	5	6.20
$ZnSO_4 \cdot 7H_2O$	13.35	8.60
KI	0.75	0.83
$Na_2MoO_4 . 2H_2O$	0.25	0.25
$CuSO_4 \cdot 2H_2O$	0.025	0.025
$CoCl_2 \cdot 6H_2O$	0.025	0.025
Vitamins		
Myo-Inositol	100	100
Nicotinamide	1	1
Pyridoxine HCl	1	1
Thiamine HCl	1	10
D-Ca Panthothenate	1	–
Choline chloride	1	–
L-ascorbic acid	2	–
Vitamin B_{12}	0.02	–
Riboflavine	0.2	–
Folic acid	0.4	–
Biotine	0.01	–
Vitamin D3	0.01	–
Vitamin A	0.01	–
p-Aminobenzoic acid	0.02	–
Other components	g/mL	g/mL
Casein hydrolysate (acid hydrolyzed)	1	–
L-glutamine	0.75	–
Maltose	85	–
Sucrose	–	30
Agarose (Sea Kem LE)	–	6

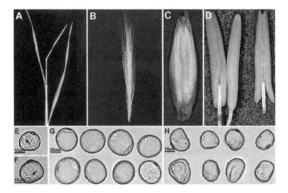

Fig. 1. Anther pre-treatment and morphology of mannitol-stressed microspores. Selection of tillers with interligule length between 4 and 6 cm (**A**). Isolated spike (**B**). Central floret (**C**). Isolated anthers from the central floret containing mid-late (**E**) to late (**F**) uninucleate microspores. Morphology of enlarged (**G**) and non-enlarged (**H**) microspores after mannitol treatment. Arrows in panels E and F indicate the position of the nucleus.

2. The tillers are surface-sterilized with 70% ethanol in the flow cabinet and placed to dry over the bottom plate of a 145- × 20-mm Petri dish (1–2 min).
3. The spikes are removed from the ensheathing leaves with the help of a set of fine-tipped forceps. The naked spikes are placed in 145- × 20-mm Petri dishes and are kept in the dark at 4 °C until anther isolation (*see* **Note 5** and **Fig. 1B**).
4. Under a binocular stereoscope, the anthers from the central floret are excised with the help of a fine forceps and placed over a glass slide (*see* **Fig. 1C and D**). The anthers are squashed in a water droplet to release the microspores. The suspension is covered with a glass cover slip and observed in a light microscope to determine the microspore developmental stage (*see* **Note 6**).
5. Only the spikes containing microspores at the ML-L uninucleate stage are used for anther isolation (*see* **Fig. 1E and F**). The anthers from the 20 middle florets are excised and placed in 0.37 M mannitol in CPW salt solution (60 anthers/mL) (*see* **Note 7**).
6. Petri dishes are sealed with Parafilm and incubated for 4 days in the dark, at 25 °C.

3.3. Microspore Isolation

1. Pour the anther suspension into an autoclaved MC2 cup of a Waring blender and add 0.37 M mannitol solution until the blades are covered with liquid. Blend the suspension for 30 s at medium power, in intermittent intervals. Remove the suspension with a 3-mL plastic pipette from the blender cup and pass it through a 110-µm nylon filter coupled to a metal cylinder to separate the microspore

suspension from the anther debris (*see* **Note 8**). Wash the blender cup with 1–3 mL of 0.37 M mannitol solution and pass it through the filter to ensure maximal microspore recovery.

2. Alternatively, for small-scale isolations, microspores can be manually isolated using a Teflon rod. To do so, pipet-off the excess of mannitol solution from the anther suspension with a 3-mL plastic pipette. Squash gently the anthers against the bottom of the Petri dish with a Teflon rod, until the solution turns milky. Add fresh mannitol solution to the microspore suspension and pass it through a 110-μm nylon filter coupled to a metal cylinder. Wash the Petri dish with 1–3 mL of mannitol solution to ensure maximal microspore recovery.

3. Collect the microspore suspension in a 100-mL beaker and transfer it with a 3-mL plastic pipet to 10-mL centrifuge tubes. Pellet the microspores by centrifugation at 100 g for 5 min at room temperature. Remove the supernatant with a glass Pasteur pipet and re-suspend the microspore pellet in 0.37 M mannitol solution up to a total volume of 10 mL. Repeat the centrifugation step once more to ensure that all anther debris are discarded in the supernatant.

3. Re-suspend the microspore pellet in 1 mL of 0.37 M mannitol solution.

4. Determine the total and the enlarged microspore density using a Fuchs–Rosenthal counting chamber (*see* **Note 9**). The morphology of enlarged and non-enlarged microspores after mannitol stress is illustrated in **Fig. 1G and H**.

5. The microspores can be cultured as a microspore suspension in liquid medium (*see* **Subheading 3.4.1**), over a solid medium layer (*see* **Subheading 3.4.2**), or immobilized in an agarose layer for time-lapse tracking of microspore-derived embryos (*see* **Subheading 3.4.3**).

3.4. Microspore Culture

3.4.1. Liquid Culture

1. Pellet the microspores isolated as described in **Subheading 3.3** by centrifugation at 100 g for 5 min at room temperature and discard the supernatant. Add medium I to obtain a final concentration of 1×10^{-4} enlarged microspores/mL.

2. Plate the microspore suspension in 60- × 15-mm or 35- × 10-mm Petri dishes (*see* **Note 10**).

3. Seal the plates with Parafilm and incubate them for 21 days in the dark at 25 °C for microspore-derived embryo development. This system allows the production of large amounts of embryos that can be used for morphological, biochemical and molecular studies. The development of microspore-derived embryos in liquid culture is illustrated in **Fig. 2A and B**.

4. After 7 days of culture, dilute them by a factor of 1.5 by adding fresh medium I.

5. After 21 days of culture, embryos ranging from 0.5 to 1 mm can be transferred to medium II and incubated at half light (600 lux, 16 h light, 25 °C).

6. After 1 week, transfer cultures to full light for regeneration (3000 lux, 16 h light, 25 °C). Microspore-derived embryo germination is illustrated in **Fig. 2D**.

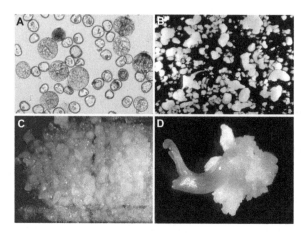

Fig. 2. Development of barley microspore-derived embryos in liquid and solid cultures. Overview of 7-day-old (**A**) and 21-day-old (**B**) microspore-derived embryos in liquid culture. Twenty-one-day-old microspore-derived embryos developed on top of a filter placed on a solid medium layer (**C**). One-week-old germinating microspore-derived embryo (**D**).

3.4.2. Solid Culture

1. Pellet the microspores isolated as described in **Subheading 3.3** by centrifugation at 100 g for 5 min at room temperature and discard the supernatant. Add medium I to obtain a final concentration of 20×10^4 enlarged microspores/mL.
2. Place a 1-cm round filter paper in the middle of a 60- × 15-mm dish with solid medium I (*see* **Note 11**).
3. Pipet 50 μL of microspore solution on top of the filter paper (*see* **Note 12**).
4. Seal the plates with Parafilm and incubate them in the dark at 25 °C for 21 days for embryo development. Embryos developed on solid medium are illustrated in **Fig. 2C**.
5. For the germination of embryos developed on solid medium, follow the same procedure as described for the liquid system (*see* **Subheading 3.4.1., steps 5 and 6**).

3.4.3. Cell Tracking

3.4.3.1. Purification of Enlarged Microspores Via a Sucrose Gradient

1. Pellet the microspores isolated as described in **Subheading 3.3** by centrifugation at 100 g for 5 min at room temperature and discard the supernatant. Add 5 mL of pre-cooled (4 °C) 15% sucrose solution (w/v) to the microspore pellet and resuspend it.

2. Add 1 mL of 0.37 M mannitol solution to the top of the sucrose layer using a 1-mL pipet connected to a Pipetboy pipettor. Dispense the mannitol very slowly and carefully, without disturbing the interface between the sucrose and the mannitol layer.

3. Centrifuge at 100 g for 10 min at 4 °C (*see* **Note 13**). After centrifugation, the enlarged microspores remain in the fraction situated at the interface between the sucrose and the mannitol, while the non-enlarged microspores are in the pellet fraction. Recover the enlarged microspore fraction using a glass Pasteur pipet and transfer it to a 10-mL centrifuge tube. Fill the tube with 0.37 M mannitol solution.

4. Pellet the microspores by centrifugation at 100 g for 5 min at room temperature and discard the supernatant. Re-suspend the microspore pellet in 1 mL of medium I.

5. Determine the enlarged microspore density using a Fuchs–Rosenthal counting chamber.

3.4.3.2. IMMOBILIZATION OF ENLARGED MICROSPORES IN AN AGAROSE LAYER

1. Place sterilized 500-μm nylon meshes pre-cut to fit into 8-well Lab-Tek II chambers (Nalgene Nunc International) with the help of a fine forceps.

2. Melt the 2% low melting type agarose stock solution in a microwave oven, and let it cool to 30 °C in a water bath.

3. Combine in a sterile Eppendorf tube the microspore suspension, the double-concentrated medium I, deionized water and the 2% low melting type agarose stock solution to a final concentration of 2×10^4 enlarged microspores/mL, 0.6% low melting type agarose in a total volume of 1 mL of medium I. Use appropriate micro-pipets to pipet the different stock solutions. For pipetting the 2% low melting type agarose stock solution, use a pipet with a cut tip. Mix well by pipetting the solution up and down with a pipet with a cut tip (*see* **Note 14**).

4. Pipet 200 μL of the mix of enlarged microspore suspension in medium I containing 0.6% (w/v) of low melting type agarose to each well, on top of the nylon mesh. Pipet the suspension carefully, taking care that a thin, even layer of immobilized microspores is formed within the mesh.

5. Wait 1–2 min until the thin layer of microspore suspension is completely solidified. Cover the thin layer of immobilized microspores with 500 μL of medium I per well.

6. Fluorescent probes, such as FDA, can be used to assay microspore viability during cell tracking. FDA is added to the 500 μL of liquid medium I at a final concentration of 1×10^{-4} μg/mL at day 0 of immobilized cultures (*see* **Note 15**).

7. Seal the chambers with Parafilm and incubate them at 25 °C in the dark for 28 days for microspore-derived embryo development (*see* **Note 16**).

8. After 7 days of culture, the cultures are diluted by a factor of 1.5 by adding 250 μL of fresh medium I to each well.

9. Embryo development can be monitored using an inverted UV microscope coupled to a digital camera. The position of each microspore is determined using the nylon mesh as a reference guide (by establishing an x and y reference

system, *see* **Fig. 3A and B**). Time-lapse tracking is done by generating digital images at appropriate intervals (depending on the purpose) by using both light and fluorescence microscopy, if fluorescent probes such as FDA are added (*see* **Note 17**). The embryogenic pathway of barley microspores, as revealed by time-lapse tracking, is illustrated in **Fig. 3C–Z**.

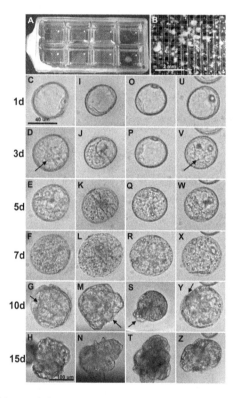

Fig. 3. Cell tracking of barley microspore-derived embryos. Overview of the immobilized culture system used for cell tracking in 8-well Lab-Tek II chambers (Nalgene Nunc International). Wells 1–6 contain immobilized cultures of enlarged microspores over 500-μm nylon meshes. Wells 7 and 8 contain control cultures where the mesh and the mesh + agarose have been omitted, respectively (**A**).Close-up of an immobilized culture after 28 days of dark incubation at 25°C. The 500-μm mesh is used to track the microspore-derived embryos in an x and y coordinates system (**B**). Typical developmental pathway of enlarged barley microspores that display embryogenic potential, as visualized by cell tracking (**C–Z**). Arrows in panels D and V indicate the central position of the nucleus (star-like morphology) and in G, M, S and Y indicate the site of exine wall rupture (opposite to the pollen germ pore).

10. After 28 days of culture, embryos ranging from 0.5 to 1 mm can be transferred to regeneration medium II, according to the procedure described for the germination of embryos developed in liquid culture (*see* **Subheading 3.4.1., step 5** and **6**).

4. Notes

1. If the osmolality and the pH of the mannitol solution are not in the range of the expected values, then do not try to adjust them. It is strongly recommended to prepare the solution again.
2. If the osmolality of the medium I is not in the range of the expected value, then do not try to adjust it. It is strongly recommended to prepare the solution again, as the osmolality of the culture media is one of the major factors to influence the androgenic response.
3. The medium I can be stored up to 1 month at 4 °C.
4. In our greenhouse conditions, we have observed that there is a good correlation between the interligule length of 4–6 cm of the first five tillers and the ML-L uninucleate stage of microspore development in the anthers for barley cv. Igri. Therefore, we advise to use only the first five tillers of the plant for optimal androgenic response. Because these morphological parameters are just an indication of the microspore stage and they vary according to the growth conditions and the genotype, a microscopic examination of squashed anthers is strongly recommended to determine the actual correlation between the interligule length, microspore stage and tiller number.
5. Tillers and isolated spikes can be kept at 4 °C up to a couple of hours before they can be further processed without any damage to microspore viability.
6. Barley microspores at the ML-L uninucleate stage are characterized by a large central vacuole and a single nucleus that is at the periphery of the cell, almost reaching the opposite side of the pollen germ pore. This stage can be visualized in anther squashes without any nuclear staining as the position of the nucleus in relation to the pollen germ pore is easily recognizable under normal light microscope (*see* **Fig. 1E and F**, nuclei are indicated by arrows).
7. As an indication, the anthers of a maximum of six spikes should be plated in a 60- × 15-mm dish, while for a 35- × 10-mm dish, it is advisable to plate the anthers of maximum one spike.
8. During microspore isolation, filtration over a 110-μm nylon mesh allows full microspore recovery, but it also allows some anther debris to pass through. This is not considered to be a problem, as the anther debris will remain in the suspension after centrifugation at 100 g and will not be collected with the microspore pellet.
9. After mannitol treatment, two morphologically distinct microspore populations can be identified: enlarged microspores that range from 40 to 60 μm diameter and have yellow/red interference of the exine wall, and non-enlarged ones that range from 20 to 30 μm diameter and display blue/black interference of the exine (*2*). Because the exine interference may vary according to the microscope

used, the exine wall colours indicated here cannot always be used to distinguish between the different classes of microspores. Therefore, the size difference can be a more accurate measure for the visual identification of enlarged and non-enlarged microspores (*see* **Fig. 1G and H**). Enlarged microspores usually constitute 20–60% of the total population. Because embryogenic microspores are among the enlarged ones *(7)*, it is crucial to perform an accurate cell counting for the correct adjustment of the enlarged cell density in order to obtain optimal androgenic response in culture. The Fuchs–Rosenthal counting chamber is used instead of other ones, because this chamber is normally used for larger cells. Other types of counting chambers are not suitable for the large size of the microspores.

10. While plating, mix the microspore suspension stock to avoid that the cells sink to the bottom. This will prevent differences in the enlarged microspore density among plates, and as a result, differences in the androgenic response.

11. By pipetting the microspores on the top of a filter paper, it is possible to obtain a dryer environment and a better medium concentration gradient for the developing microspores. Using no filter and pipetting the microspores directly on the solid medium leads to a wetter environment for the microspores, and there is a risk of having medium all around the microspores instead of only from beneath. The best embryo formation is found with the use of filter paper.

12. The high density of microspores on a small square was tested to be the best way to perform this type of culture.

13. Separation of enlarged microspores by a sucrose gradient can be performed at room temperature; however, a better separation is obtained when the centrifugation step is performed at 4 °C.

14. While mixing the microspore suspension for immobilization, always add the 2% agarose stock solution at last with a pipet with a cut tip. This will prevent formation of air bubbles. After adding the agarose, work fast as the solution will be solidified within a few minutes. Always perform control cultures without the nylon mesh and the agarose.

15. If FDA is added to the immobilized cultures, it is strongly recommended to use it from a FDA stock prepared with ethanol rather than acetone. Use the lowest concentration possible.

16. Embryo development is slower in immobilized cultures than in liquid; however, the efficiency is comparable to the liquid system. In immobilized cultures, up to 10% of the enlarged microspores develop into embryos.

17. In our hands, an Olympus microscope coupled to a Nikon DXM 1200 digital camera has been used. Dry lenses are preferred. If fluorescent probes are used, make sure that the UV light is properly adjusted and that appropriate neutral density filters are used.

References

1. Wang M, van Bergen S, van Duijn B. Insights into a key developmental switch and its importance for efficient plant breeding. Plant Physiol 2000;124:523–530.

2. Hoekstra S, van Zijderveld MH, Louwerse JD, Heidekamp F, van der Mark F. Anther and microspore culture of *Hordeum vulgare* L. cv. Igri. Plant Sci 1992;86:89–96.

3. Jähne-Gärtner A, Lörz H. Protocols for anther and microspore culture of barley. In: Hall RD, ed. Methods in Molecular Biology. Totowa, NJ: Humana, 1999:269–279.

4. Maraschin SF, Lamers GEM, de Pater BS, Spaink HP, Wang M. 14-3-3 isoforms and pattern formation during barley microspore embryogenesis. J Exp Bot 2003;51:1033–1043.

5. Maraschin SF, Caspers M, Potokina E, Wülfert F, Graner A, Spaink HP, Wang M. cDNA array analysis of stress-induced gene expression in barley androgenesis. Physiol Plant 2006;127:535–550.

6. Vrienten PL, Nakamura T, Kasha KJ. Characterization of cDNAs expressed in the early stages of microspore embryogenesis in barley (*Hordeum vulgare*) L. Plant Mol Biol 1999; 41:455–463.

7. Maraschin SF, Vennik M, Lamers GEM, Spaink HP, Wang M. Time-lapse tracking of barley androgenesis reveals position-determined cell death within pro-embryos. Planta 2005;220:531–540.

8. Kumlehn J, Lörz H. Monitoring sporophytic development of individual microspores of barley (*Hordeum vulgare* L.). In: Clement C, Pacini E, Audran JC, eds. Anther and Pollen: From Biology to Biotechnology. Berlin Heidelberg: Springer-Verlag, 1999:183–189.

9. Bolik M, Koop HU. Identification of embryogenic microspores of barley (*Hordeum vulgare* L.) by individual selection and culture and their potential for transformation by microinjection. Protoplasma 1991;162:61–68.

10. Maraschin SF, Gaussand G, Pulido A, Olmedilla A, Lamers GEM, Korthout H, Spaink HP, Wang M. Programmed cell death during the transition from multi-cellular structures to globular embryos in barley androgenesis. Planta 2005;221: 459–470.

7

Isolation of Embryo-Specific Mutants in *Arabidopsis*

Plant Transformation

Nai-You Liu, Zhi-Feng Zhang, and Wei-Cai Yang

Summary

As a model system for efficient analysis of plant gene function, *Arabidopsis* can easily be transformed by *Agrobacterium*-mediated gene transfer. Among the various transformation methods developed during the past two decades, the vacuum infiltration transformation and the floral dip method are most widely used. Compared with conventional plant transformation methods that involve tissue culture and plant regeneration, vacuum infiltration and the floral dip method require minimal time and labor; furthermore, these two high-throughput transformation procedures can also be used in insertional mutagenesis, one of the most powerful tools for studying gene function. Here, we outline a detailed description with various fine-tuned steps for these two *Agrobacterium*-mediated floral transformation methods in *Arabidopsis*.

Key Words: *Arabidopsis*; plant transformation; *Agrobacterium tumefaciens*; vacuum infiltration; floral dip.

1. Introduction

Plant transformation is the introduction of foreign DNA into plant cells or tissues, and then a whole genetically engineered plant is regenerated. In *Arabidopsis* , plant transformation can easily be achieved by *Agrobacterium*-mediated gene transfer, which has been widely used for studying gene function. The gene of interest or a engineered transposable element are first cloned into the Transfer-DNA (T-DNA) region of a binary vector, the latter then is introduced into *Agrobacterium tumefaciens*, which help to deliver and integrate the

From: *Methods in Molecular Biology, vol. 427: Plant Embryogenesis*
Edited by: M. F. Suárez and P. V. Bozhkov © Humana Press, Totowa, NJ

T-DNA into the plant genome *(1,2)*. Both T-DNA and transposable element have been used successfully as biological mutagens in *Arabidopsis* and rice and offer a unique tool for both forward and reverse genetic approaches *(1,3–6)*.

Traditional plant transformation methods typically involve multiple steps including initiation of transformation-competent plant cells or tissues, DNA delivery to plant cells by *A. tumefaciens* or by the biolistic method, selection of transformed cells that have stably integrated foreign DNA, and regeneration of transformed cells or tissues into transgenic plants *(7,8)*. Because they require tissue culture and plant regeneration steps, these methods are painstaking and time consuming and require expertise and relatively expensive laboratory facilities.

For *Arabidopsis*, the first *"in-planta"* transformation method that avoids tissue culture and somaclonal variations was described by Feldmann and Marks in 1987, where seeds were imbibed in the presence of *Agrobacterium* *(9)*. In 1990, Hong-Gil Nam *(10)* and colleagues reported another whole plant transformation procedure in which young inflorescences were cut off and the wounded surfaces were inoculated with *Agrobacterium*. However, due to the low frequency of transformation, these methods are difficult to reproduce. In 1993, Bechtold et al. *(11)* presented the infiltration method. In this protocol, adult plants at the reproductive stage are vacuum-infiltrated in *Agrobacterium* suspension so that reproductive organs are invaded by the bacteria. Combining with its higher transformation frequency (over 0.4% of the seeds) and its reproducibility and simplicity, the vacuum-infiltration method has been adopted widely as a random mutagenesis method in *Arabidopsis*. At the end of the last century, Clough and Bent showed that the vacuum-aided infiltration of inflorescences could be substituted by the use of a surfactant Silwet L-77 *(12)*. Using this floral dip method, a transformation frequency of at least 1% can be routinely obtained *(7)*. Now it is a routine method in *Arabidopsis* transformation. This chapter describes a detailed protocol for *Agrobacterium*-mediated vacuum infiltration and the floral dip method in *Arabidopsis* .

2. Materials

2.1. Plant Materials

Arabidopsis thaliana ecotypes Columbia (Col-0), Landsberg *erecta* (Ler-0), Wassilevskija (WS-0), Niederzenz (Nd-0), Noordwijk (Nok-0), Nossen (No-0), Tsu (Tsu-0), Enkheim (En-2), Cape Verdi Islands (Cvi-0), C24 (C24) (*see* **Note 1**).

Table 1
List of T-DNA Vectors *(26)*

Vector	Unique cloning sites	LacZ selection	Resistance in	
			bacteria	plants
pBIN19	9	No	kanamycin (Kan)	Kan
pCAMBIA series	Variable	Yes (not all)	chloramphenicol (Chlor), Kan	hygromycin (Hyg), Kan
pCGN series	5	Yes	gentamycin (Gent)	Kan
pJJ/pSLJ series	5–11	Yes	tetracycline	herbicide (Bar), streptomycin, Kan, Hyg
pPZP series	9	Yes	Chlor, Spec	Kan, Gent
pGreen series	18	Yes	Kan	Sulfonamide, Bar, Kan, Hyg

2.2. Agrobacterium Strains and Vectors

1. *A. tumefaciens* strains: AGL-1 *(13)*, GV3101 (pMP90) and GV3101 (pMP90RK) *(14)*, LBA4404 *(15)*, EHA105 *(16)*, ABI *(17,18)*.
2. Vectors: pBIN19 *(19)*, pCAMBIA series (http://www.cambia.org.au/), pCGN series *(20)*, pJJ/pSLJ series *(21)*, pPZP series *(22)*, pGreen series *(23)* (*see* **Note 2** and **Table 1**).

2.3. Lab Equipment and Supplies

1. Air-conditioned *Arabidopsis* growth chambers or greenhouse under a 16-h light /8-h dark cycle at 22 ± 2 °C.
2. Petri dishes: e.g., 90 mm × 12 mm with 98 mm cover (Falcon).
3. Vacuum apparatus.
4. Minisart Syringe Filter with 0.2-μm membranes (Sartorius).
5. Greenhouse materials

   a. Plant pots: e.g., 4 in. × 4 in. square pots.
   b. Flower pot holders (plastic): e.g., 40 cm × 20 cm ×5 cm plastic trays.
   c. Plastic domes.
   d. Soil or potting composite.

2.4. Medium

1. *Agrobacterium* culture medium

   a. Luria-Bertani (LB) medium (1 L): 10 g Bacto-tryptone (OXOID), 5 g Bacto-Yeast extract (OXOID), 10 g NaCl, pH 7.0; pH is adjusted with 1 M NaOH. The medium is sterilized by autoclaving at 121 °C for 15 min (*see* **Note 3**).

b. YEP liquid medium (1 L): 10 g Peptone (OXOID), 10 g Bacto-Yeast extract (OXOID), 5 g NaCl, pH 7.0; pH is adjusted with 1 M NaOH. The medium is sterilized by autoclaving at 121 °C for 15 min.

2. Transformation media

a. Infiltration medium (1 L): 4.33 g Murashige & Skoog Basal Salts Mixture (Sigma M5524) *(24,25)*, 50 g sucrose, 50 μL Silwet L-77 (Lehle Seeds), pH 5.8; freshly prepared.

b. Floral dipping medium (1 L): 4.33 g Murashige & Skoog Basal Salts Mixture (Sigma M5524), 50 g sucrose, 200 μL Silwet L-77 (Lehle Seeds), pH 5.8; freshly prepared.

c. Murashige & Skoog (MS) medium (1 L): 4.33 g Murashige & Skoog Basal Salts Mixture (Sigma M5524), 10 g sucrose, pH 5.8, autoclave at 121 °C for 15 min.

d. Nutrient solution: 40 mL 25× B5 Macronutrients, 1 mL 1000× B5 Micronutrients, 10 mL 100× Fe-EDTA, and then fill up to 1 L.

   i. 25× B5 Macronutrients (1 L): 4.24 g $NaH_2PO_4 \cdot 2H_2O$, 62.5 g KNO_3, 3.35 g $(NH_4)SO_4$, 6.25 g $MgSO_4$ $7H_2O$, 2.84 g $CaCl_2$.

   ii. 1000× B5 Micronutrients (1 L): 10 g $MnSO_4 \cdot H_2O$, 3 g H_3BO_3, 2 g $ZnSO_4 \cdot 7H_2O$, 0.25 g $NaMoO_3$, 0.025 g $CuSO_4 \cdot 5H_2O$, 0.025 g $CoCl_2 \cdot 6H_2O$, 0.75 g KI.

   iii. 100× Fe-EDTA (200 mL): 0.556 g $FeSO_4 \cdot 5H_2O$, 0.746 g EDTA-Na.

2.5. Selection Plates

Add 8 g agar to 1 L MS medium and autoclave at 121 °C for 15 min. Then add the appropriate selection agents such as antibiotics and herbicide (filter-sterilized) when the medium cool to about 50–60 °C, mix thoroughly and dispense into Petri dishes, solidify at room temperature (*see* **Note 4**).

2.6. Reagents for Seed Surface Sterilization

Sterile distilled water; 70% (vol/vol) ethanol; 20% (vol/vol) bleach (contains about 0.6–1% NaClO) + 1% (vol/vol) Tween-20 (Sigma) (not essential); freshly prepared.

3. Methods
3.1. Arabidopsis Cultivation

1. Seed germination and seedling growth. Here, we present two methods for germinating seeds and growing seedlings.

a. Cultivate the plants directly in soil. Suspend dry *Arabidopsis* seeds in 0.05–0.1% agar or sterile water and then keep them in darkness at 4 °C for 2–3 days (*see* **Note 5**). Prepare pots (e.g., 4 in. × 4 in.) with wet soil mixtures (perlite and vermiculite 1:1 mixed). Pipette about 5–10 seeds directly onto the soil mixture in each pot. Alternatively, one can use wet toothpick tip to transfer the seeds to soil. Cover the pots with a transparent plastic membrane to maintain high humidity for the first 10 days.

b. For many particular applications, such as determining germination frequency, selecting for transformed plants, and quickly preparing plants for transformation, the following method is more desirable. Seeds surface-sterilization: Sterilize seeds first by treating them with 20% (vol/vol) bleach + 1% (vol/vol) Tween-20 for 5 min, and then with 70% ethanol for another 5 min followed by three to four times washing with sterile water. Then resuspend and spread the seeds onto MS medium plates supplemented with appropriate selection agents such as antibiotics or herbicide. Keep the plates in darkness at 4 °C for 2–3 days (*see* **Note 5**). Transfer the plates to greenhouse under long-day conditions (16-h light/8-h dark, 22 ± 2 °C) for about 10–14 days. And then transfer the seedlings to wet soil.

2. Grow plants in a growth chamber or a greenhouse under short-day conditions (e.g., 13-h light) for 3–4 weeks until the bolts emerge and produce floral inflorescences. Then move them to long-day conditions with 16-h day photoperiod. Water the plants two or three times with nutrient solution during flowering (*see* **Note 6**).

3.2. Agrobacterium Culture and Preparation

1. Three days before plant transformation, prepare a preculture of *A. tumefaciens* strain carrying the gene of interest in a binary vector by inoculating 5 mL of LB liquid medium containing the appropriate antibiotics with a single *Agrobacterium* colony (*see* **Note 7**).

2. Incubate at 28 °C with vigorous shaking (250 rpm) for 1–2 days.

3. Use the preculture to inoculate 500 mL LB medium with the appropriate antibiotics.

4. Grow at 28 °C with vigorous shaking and good aeration for 1–2 days until the OD (600 nm) reaches 1.2–2.0 (*see* **Note 8**).

5. Collect *Agrobacterium* cells by centrifuging at 4000 g for 15 min at room temperature.

6. Resuspend the bacteria pellet gently in 1 L transformation medium (OD ≈0.8) (*see* **Note 9**).

3.3. Plant Transformation

1. Vacuum infiltration (*11*) (*see* **Note 10**).

a. Transfer the *Agrobacterium* suspension to four 250-mL beakers (*see* **Note 11**).

b. Invert a pot of plants and submerge the aerial parts of the plants including all the bolts into the *Agrobacterium* cell suspension (*see* **Fig. 1 B and C** and **Note 12**).

c. Place the beakers inside a bell jar and apply 10⁴Pa of vacuum pressure for 10–15 min (*see* **Fig. 1D**). Gently break the vacuum (*see* **Note 13**).

Fig. 1. Vacuum infiltration and floral dip transformation of *Arabidopsis*. (**A**) The optimal stage for transformation is when the plants have just started to flower and the first siliques are being formed *(27)*. (**B**) Cover soil with a central-holed plastic plate to prevent loss of soil when the plants are inverted. (**C**) Invert the plants and submerge the aerial parts into an *Agrobacterium* suspension. For floral dip method, just dip the aerial parts of plants in the *Agrobacterium* suspension for 5–10 s. (**D**) Vacuum at 10^4 Pa for 10–15 min (*see* **Note 10**). (**E**) Cover the transformed plants with a transparent dome to maintain high humidity for 24 h. (**F**) Remove the cover and grow the plants in greenhouse until maturity. Selection of transformants on MS plates containing either kanamycin (+carbenicillin) (**G**) or hygromycin (+carbenicillin) (**H**).

    d. Remove pots from beakers and lay them down on their side in a plastic tray. Cover with a transparent dome to maintain high humidity for 24 h (*see* **Fig. 1 E** and **Note 14**).

    e. Remove the cover and return the plants to their normal growing conditions until maturity and then let the plants dry by reducing watering progressively (*see* **Fig. 1F**).

    f. After 4–6 weeks, harvest seeds and let them air dry. Dried seeds can be stored at room temperature for a few months or at 4 °C for 1–3 years.

2. Floral Dip

    a. Put the *Agrobacterium* suspension in an appropriate vessel.

    b. Dip the developing inflorescence into the *Agrobacterium* suspension for 5–10 s with gentle agitation (*see* **Fig. 1B and C** and **Note 15**).

    c. Cover the dipped plants with a transparent dome for 24 h to maintain high humidity (*see* **Fig. 1E** and **Note 14**).

    d. Proceed through the **steps 1d, 1e and 1f** described in 'Vacuum Infiltration' above.

3.4. Screening of the Transformants

1. Prepare the selection plates containing carbenicillin and appropriate selection agents (antibiotics or herbicide, *see* **Subheading 2.5**) (*see* **Note 16**).
2. Seed sterilization and seedling cultivation (*see* **Subheading 3.1**).
3. Check the plates for the transformants, which are visible as green seedlings with true leaves and long roots (*see* **Fig. 1G and H** and **Note 17**).
4. Transfer the potential transformants to a new selection plate and allow them to grow 7–14 days to confirm that they are true transgenic lines *(7)*.
5. Carefully transplant the plantlets from the selection plates onto wet soil. Cover them with a plastic dome to maintain high humidity for 2 days.
6. Grow the plants in greenhouse under 16-h light/8-h dark cycle at 22 ± 2 °C.

4. Notes

1. The methods we present here work well for the majority of *Arabidopsis thaliana* ecotypes.
2. Considerations when choosing a binary vector include its size and copy number, selection marker in bacteria and plants, availability of LacZ selection and unique restriction sites for cloning (*see* **Table 1**).
3. Unless stated otherwise, all solutions should be prepared in distilled water and should be stored at room temperature.
4. Antibiotics and herbicide final concentrations: kanamycin, 25–50 mg/L; hygromycin, 5–30 mg/L; carbenicillin, 100 mg/L.
5. Seed dormancy can be broken by keeping the seeds at 4 °C for 2–3 days before germination. This also results in synchronous seed germination.
6. Plant health is a major factor for a successful transformation. We find that healthier plants can be achieved by growing plants first under relatively short-day condition (e.g., 13 h) and then under long-day condition. Watering the plants

two or three times with nutrient solution during flowering is also recommended. The optimal stage (*see* **Fig. 1A**) for transformation is when the plants have just started to flower and the first siliques are being formed *(27)*. Furthermore, in order to produce plants with more secondary inflorescences, we often cut off the tip of the first primary bolt.

7. Three means to confirm the presence of desired construct: by digestion with appropriate restriction enzymes, by polymerase chain reaction or by sequencing the gene to be transferred *(7)*.

8. It takes about 20 h for *Agrobacterium* strain such as MP5-1 to reach the desired OD, but it may take longer for other strains. We find that YEP liquid medium gives a higher *Agrobacterium* yield.

9. Surfactant Silwet L-77 is critical for a successful transformation. But high concentrations might be toxic to plants. We recommend the concentrations mentioned in **Subheading 2.4.**

10. For transformation of *Arabidopsis* ecotype Ler-0, vacuum infiltration is preferred *(7)*, because lower frequency is obtained by other methods *(12)*.

11. You can use the same suspension for three times.

12. The soil should be well watered 1 day before transformation, otherwise it will absorb much of the *Agrobacterium* suspension. Plants with numerous immature floral buds and fewer siliques when inoculated give the highest transformation efficiency *(12)*.

13. Good infiltration is essential for a successful transformation. The vacuum time and pressure vary in different laboratories. To vacuum the plants twice with a 7-day interval is also recommended.

14. Avoid high illumination and water condensation at this step. Remember do not completely cover the plants.

15. Make sure that a film of *Agrobacterium* suspension coats the surfaces of all buds on the inflorescences. Alternatively, you can apply the suspension to the floral buds either by using a micropipettor (Pipetman) or by spray, which is called floral spray transformation *(28)*. These two methods both work well for *Arabidopsis*. Also, dipping or spraying the plants at a 7-day interval is helpful for obtaining higher transformation frequency.

16. Carbenicillin is used to prevent possible *Agrobacteria* contamination *(7)*. The plates should be dried well before use.

17. For kanamycin (+carbenicillin) selection, untransformed seedlings have yellow cotyledons and short roots. For hygromycin (+carbenicillin) selection, untransformed seedlings have green cotyledons, but they can not form true leaves and their roots are short.

References

1. Page DR, Grossniklaus U. The art and design of genetic screens: *Arabidopsis thaliana* . Nat Rev Genet 2002;3:124–136.
2. Pan XK, Li Y, Stein L. Site preferences of insertional mutagenesis agents in *Arabidopsis*. Plant Physiol 2005;137:168–175.

3. Dubois P, Cutler S, Belzile FJ. Regional insertional mutagenesis on chromosome III of *Arabidopsis thaliana* using the maize *Ac* element. Plant J 1998;13:141–151.
4. Krysan PJ, Young JC, Sussman MR. T-DNA as an insertional mutagen in *Arabidopsis*. Plant Cell 1999;11:2283–2290.
5. Nishal B, Tantikanjana T, Sundaresan V. An inducible targeted tagging system for localized saturation mutagenesis in *Arabidopsis*. Plant Physiol 2005;137:3–12.
6. Jeon J-S, Lee S, Jung K-H, Jun S-H, Jeong D-H, Lee J, Kim C, Jang S, Lee S, Yang S, Yang K, Nam J, An K, Han M-J, Sung R-J, Choi H-S, Yu J-H, Choi J-H, Cho S-Y, Cha S-S, Kim S-I, An G. T-DNA insertional mutagenesis for functional genomics in rice. Plant J 2000;22:561–570.
7. Zhang X, Henriques R, Lin S-S, Niu Q-W, Chua N-H. *Agrobacterium*-mediated transformation of *Arabidopsis thaliana* using the floral dip method. Nat Protoc 2006;1:641–646.
8. Christou P. Transformation technology. Trends Plant Sci 1996;1:423–431.
9. Feldmann KA, Marks MD. *Agrobacterium*-mediated transformation of germinating seeds of *Arabidopsis thaliana*: a non-tissue culture approach. Mol Gen Genet 1987;208:1–9.
10. Chang SS, Park SK, Nam HG. Transformation of *Arabidopsis* by *Agrobacterium* inoculation on wounds. Plant J 1990;5:551–558.
11. Bechtold N, Ellis J, Pelletier G. *In planta Agrobacterium* mediated gene transfer by infiltration of adult *Arabidopsis thaliana* plants. C R Acad Sci Paris, Life Sci 1993;316:1194–1199.
12. Clough SJ, Bent AF. Floral dip: a simplified method for *Agrobacterium*-mediated transformation of *Arabidopsis thaliana*. Plant J 1998;16:735–743.
13. Lazo GR, Stein PA, Ludwig RA. A DNA transformation-competent *Arabidopsis* genomic library in *Agrobacterium*. Biotechnology (NY) 1991;9:963–967.
14. Koncz C, Schell J. The promoter of TL-DNA gene 5 controls the tissue-specific expression of chimeric genes carried by a novel type of *Agrobacterium* binary vector. Mol Gen Genet 1986;204:383–396.
15. Ooms G, Regensburg-Tuink TJG, Hofker MH, Hoekema H, Hooykaas PJJ, Schilperoort RA. Studies on the structure of cointegrates between octopine and nopaline Ti-plasmids and their tumor-inducing properties. Plant Mol Biol 1982;1:265–276.
16. Hood EE, Gelvin SB, Melchers LS, Hoekema A. New *Agrobacterium* helper plasmids for gene transfer to plants. Transgenic Res 1993;2:208–218.
17. Koncz C, Kreuzalerl F, Kalmanl Z, Schell J. A simple method to transfer, integrate and study expression of foreign genes, such as chicken ovalbumin and α-actin in plant tumors. EMBO J 1984;3:1029–1037.
18. Cheng M, Fry JE, Pang S, Zhou H, Hironaka CM, Duncan DR, Conner TW, Wan Y. Genetic transformation of wheat mediated by *Agrobacterium tumefaciens*. Plant Physiol 1997;115:971–980.
19. Bevan M. Binary *Agrobacterium* vectors for plant transformation. Nucleic Acids Res 1984;12:8711–8721.
20. McBride KE, Summerfelt KR. Improved binary vectors for *Agrobacterium*-mediated plant transformation. Plant Mol Biol 1990;14:269–276.

21. Jones JD, Shlumukov L, Carland F, English J, Scofield SR, Bishop GJ, Harrison K. Effective vectors for transformation, expression of heterologous genes, and assaying transposon excision in transgenic plants. Transgenic Res 1992;1:285–297.
22. Hajdukiewicz P, Svab Z, Maliga P. The small, versatile pPZP family of *Agrobacterium* binary vectors for plant transformation. Plant Mol Biol 1994;25:989–994.
23. Hellens RP, Edwards EA, Leyland NR, Bean S, Mullineaux PM. pGreen: a versatile and flexible binary Ti vector for *Agrobacterium*-mediated plant transformation. Plant Mol Biol 2000;42:819–832.
24. Murashige T, Skoog F. A revised medium for rapid growth and bioassays with tobacco tissue culture. Physiol Plant 1962;15:473–497.
25. Estelle MA, Somerville C. Auxin-resistant mutants of *Arabidopsis thaliana* with an altered morphology. Mol Gen Genet 1987;206:200–206.
26. Weigel D, Glazebrook J. How to transform *Arabidopsis*. In: Weigel D, Glazebrook J, ed. Arabidopsis, A Laboratory Manual. New York: Cold Spring Harbor Laboratory Press, Cold Spring Harbor, 2002:119–140.
27. Bent AF. *Arabidopsis in planta* transformation: uses, mechanisms, and prospects for transformation of other species. Plant Physiol 2000;124:1540–1547.
28. Chung MH, Chen MK, Pan SM. Floral spray transformation can efficiently generate *Arabidopsis* transgenic plants. Transgenic Res 2000;9:471–476.

8

Isolation of Embryo-Specific Mutants in *Arabidopsis*
Genetic and Phenotypic Analysis

Nai-You Liu, Zhi-Feng Zhang, and Wei-Cai Yang

Summary

To perform an effective genetic screen for embryo-specific mutants is a prerequisite for understanding the molecular mechanisms that control *Arabidopsis* embryo development. Mutagenesis based on either T-DNA or transposon insertion has been successfully used in identifying embryonic mutants. We present here a typical genetic screen for putative embryonic mutants based on distorted Mendelian segregation ratio (2:1) and reduced seed set. It is advisable to examine whether the mutation also affects gametophytic functions by performing reciprocal crosses between wild type and the mutant. We also provide detailed explanations on the whole-mount clearing method, a simple but effective method for phenotypic analysis of mutant embryos blocked in certain steps during the process necessary for embryo viability and development.

Key Words: Genetic screen; Mendelian segregation; embryo; reciprocal crosses; whole-mount clearing method.

1. Introduction

Forward genetic screen, by which genetic variation is artificially induced and mutagenized plants are screened for phenotypes of interest *(1)*, is a powerful strategy for identifying genes that are involved in particular biological processes. For *Arabidopsis*, mutagens commonly used include ethylmethane sulfonate, fast neutron radiation, and insertion elements such as T-DNA and transposons. The advantage of T-DNA or transposon mutagenesis is that the known DNA insertion element can be used as a "tag," by which the *Arabidopsis* DNA sequences flanking the insertion site can be easily identified, and hence the

From: *Methods in Molecular Biology, vol. 427: Plant Embryogenesis*
Edited by: M. F. Suárez and P. V. Bozhkov © Humana Press, Totowa, NJ

interrupted gene *(2)*. To date, extensive mutagenization programs for embryo-defective mutants *(emb)* mutants have been performed in *Arabidopsis (3–7)*, and a minimal set of 750 nonredundant *Arabidopsis* genes are predicted to be essential for embryogenesis *(6)*. Here, we describe a typical genetic screen for putative embryo mutants based on distorted Mendelian segregation ratio (2:1) and reduced seed set *(1,8)* from large collections of kanamycin (Kan)-marked T-DNA or *Dissociation (Ds)* transposon insertional lines *(9,10)*. Embryo mutants are generally maintained as heterozygotes that produce 25% aborted seeds after self-pollination *(11)*. That is, the siliques of the mutant are expected to contain normal and defective embryos in a 3:1 ratio. Because reduced fertility can also be caused by other reasons such as inappropriate environmental conditions, a second screen step to confirm the mutation is necessary. Distorted Kan marker segregation ratio can indicate embryonic or gemetophytic lethality. For an idealized mutation, a 2:1 segregation ratio of Kan resistant (Kan^R) to sensitive seedlings is expected *(1)*. It is advisable to examine whether the mutation also affects gametophytic functions by performing reciprocal crosses between wild type and the mutant *(8)*. Phenotype of embryo mutants can be studied by classical and confocal microscopy, histology, and marker gene analysis *(12)*. We provide here a detailed protocol on the whole-mount clearing method *(13,14)*, a simple but effective method for phenotypic analysis of mutant embryos blocked in certain steps during the process necessary for embryo viability and development *(15)*.

2. Materials

2.1. Plant Materials

1. T-DNA and transposon insertional lines from two main stock centers: the *Arabidopsis* Biological Resource Center (ABRC) at Ohio State University (http://www.biosci.ohio-state.edu/pcmb/Facilities/abrc/abrchome.htm) and the Nottingham *Arabidopsis* Stock Center (NASC) in the United Kingdom (http://nasc.nott.ac.uk/).
2. *Ds* transposon insertional lines *(9,10)*.
3. *Arabidopsis thaliana* ecotypes Columbia (Col-0) and Landsberg *erecta* (Ler-0).
4. *Ds* insertional line *fy60* (*see* **Note 1**).

2.2. Equipments

1. Air-conditioned *Arabidopsis* growth chambers or greenhouse under a 16-h light/8-h dark cycle at 22 ± 2 °C.
2. Dissecting microscope.
3. A microscope, such as Zeiss Axioskop II or Leica DMRB, equipped with differential interference contrast (DIC) optics and image capturing system.
4. Microscope cover glasses (Sigma CLS286518).

5. Tweezers: needle sharp, Style no. 5, (e.g., T4537-1EA, Sigma-Aldrich).
6. Hypodermic needles: 1/2cc U-100 insulin injection syringe.
7. Minisart Syringe Filter with 0.2-μm membranes (Sartorius).
8. Double-sided tape.
9. Microscope slides (Sigma S8902).

2.3. Reagents

1. Murashige & Skoog (MS) medium (1 L): Dissolve 4.33 g Murashige & Skoog Basal Salts Mixture (Sigma M5524) *(16,17)* and 10 g sucrose in 800 mL distilled water, adjust to pH 5.8, fill up to 1 L and autoclave at 121 °C for 15 min (*see* **Note 2**).
2. Selection plates: Add 8 g agar to 1 L MS medium and autoclave at 121 °C for 15 min. When the medium cool to about 50–60 °C, add 1 mL of 50 mg/mL Kan stock solution (filter-sterilized and stored at –20 °C), to a final concentration of 50 mg/L. Pour the plates and let them solidify. Store selection plates at 4 °C.
3. Seed surface sterilization: Sterile distilled water; 70% (vol/vol) ethanol; 20% (v/v) bleach (contains about 0.6–1% NaClO) + 1% (vol/vol) Tween-20 (Sigma) (not essential), freshly prepared.
4. Nutrient solution: 40 mL 25× B5 macronutrients, 1 mL 1000× B5 micronutrients, and 10 mL 100× Fe-EDTA, to 800 mL distilled water, adjust pH to 5.8, and then fill up to 1 L. Freshly made.

   a. 25× B5 macronutrients (1 L): 4.24 g $NaH_2PO_4 \cdot 2H_2O$, 62.5 g KNO_3, 3.35 g $(NH_4)SO_4$, 6.25 g $MgSO_4 \cdot 7H_2O$, 2.84 g $CaCl_2$.
   b. 1000× B5 Micronutrients (1 L): 10 g $MnSO_4 \cdot H_2O$, 3 g H_3BO_3, 2 g $ZnSO_4 \cdot 7H_2O$, 0.25 g $NaMoO_3$, 0.025 g $CuSO_4 \cdot 5H_2O$, 0.025 g $CoCl_2 \cdot 6H_2O$, 0.75 g KI.
   c. 100× Fe-EDTA (200 mL): 0.556 g $FeSO_4 \cdot 5H_2O$, 0.746 g EDTA-Na.

5. Herr's solution: Lactic acid : chloral hydrate:phenol : clove oil : xylene (2:2:2:2:1, w/w) *(13,14)*. Remember to wear safety glasses, gloves, suitable respiratory equipment, and have good ventilation as phenol and xylene are toxic.

3. Methods

3.1. Genetic Screen for Putative Embryo Mutants

3.1.1. Segregation Ratio Analysis of Kanamycin Selection Marker

1. Sterilize progeny seeds from T-DNA or *Ds* insertion lines harboring a Kan[R] marker (*see* **Fig. 1A**) by treating them with 20% (vol/vol) bleach + 1% (vol/vol) Tween-20 for 5 min, and then with 70% ethanol for another 5 min followed by washing the seeds three to four times with sterile water.
2. Resuspend the seeds in sterile water and spread onto selection plates with a pipette (*see* **Note 3**).

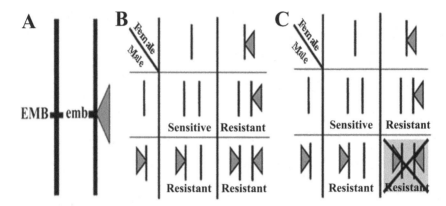

Fig. 1. Segregation analysis of kanamycin (Kan) selection marker. (**A**) Insertional mutagenesis of *Arabidopsis* using T-DNA or the *Ac/Ds* system. An insertional mutagen (T-DNA or *Ds*) disrupts an *Arabidopsis* embryo-defective gene (EMB) essential for embryo development and introduces a dominant marker that confers resistance to kanamycin (Kan^R). (**B**) A drawing showing typical Mendenlian segregation (3:1) of the insertion. (**C**) Distorted segregation ratio of the dominant Kan^R marker indicates embryo lethality. For an idealized mutation with full penetrance, a 2:1 segregation ratio of resistant to sensitive seedlings is expected.

3. Keep the plates at 4 °C in darkness for 2–3 days (*see* **Note 4**).
4. Transfer the plates to greenhouse under long-day conditions with 16-h light/8-h dark cycle at 22±2 °C and grow for about 10–14 days.
5. Carefully score the Kan^R and Kan-sensitive (Kan^S) seedlings on the plate under a dissecting microscope (*see* **Note 5**). The Kan^S seedlings have yellow cotyledons and short roots, whereas the Kan^R seedlings have green cotyledons and long roots. For a putative embryo mutant, the Kan^R : Kan^S segregation ratio is expected to be 2:1 compared with 3:1 ratio for a recessive sporophytic mutant (*see* **Note 6** and **Fig. 1B and C**) *(8)*.
6. Transfer the Kan^R seedlings to soil for further phenotypic analysis and crossing.

3.1.2. Screen for Reduced Seed Set

1. Transfer the Kan^R seedlings to wet soil.
2. Cover the plants with a transparent dome to maintain high humidity for 1–2 days (*see* **Note 7**).
3. Grow plants in a growth chamber or a greenhouse under long-day conditions. Water the plants two or three times with nutrient solution during flowering.
4. Dissect mature siliques with a syringe needle. We often fix the silique onto a double-sided tape on a glass slide and count the aberrant seeds and normal seeds under a binocular or dissecting scope. In wild-type siliques, full seed set should

be observed. Whereas in mutant plants grown under the same conditions as wild type, approximately 25% of the ovules are expected to be small, shrunken, and aborted (*see* **Note 8**) *(8)*.

3.2. Reciprocal Crosses

To investigate whether the T-DNA or *Ds* insertion also affected gametophytic functions, reciprocal crosses between the mutant and wild-type plants should be performed *(8,18)*. If the mutation does not affect gametophytic function, the F1 progenies should segregate 1:1 for Kan^R and Kan^S. A detailed crossing procedure is as follows.

1. Select a healthy mutant plant as female and a wild-type plant as male. Label properly with a waterproof marker pen.
2. Select mutant flowers whose petals just turn white before anthesis as female. A healthy plant often has two to three flowers on the main shoot that are suitable for crossing each day (*see* **Note 9**). Remove any self-pollinated siliques (*see* **Note 10**).
3. Emasculate the flower by carefully removing the sepals, petals, and stamens with a pair of tweezers, and leaving the stigma intact. Make sure that all the stamens are removed and the pollen is not shed onto the stigma under a microscope (*see* **Note 11**). Tube the plant with transparent plastic sheet to prevent drying of the stigma and cross-pollination.
4. Carefully brush the stigma surface of the emasculated flower in 18–24 h with dehiscent anthers of a wild-type flower under a binocular (*see* **Note 11**). Mark the pollinated flower with a color thread (optional).
5. Pollinate more flowers as described in **steps 1–4** until the desired number of seeds is achieved. A successful cross should give rise to 35–50 seeds per silique.
6. In 3–4 weeks, the seeds should be ready for harvest. Collect the siliques as soon as possible when they turn yellow.
7. Dry the siliques for 7 days at 37 °C or 14 days at room temperature. F1 seeds can be stored at room temperature for a few months or 4 °C for 1–3 years.
8. Check F1 seeds segregation ratio as described in **Subheading 3.1.1.** When heterozygous mutant plants are used as either egg or pollen donors, the F1 progeny are expected to segregate at a Kan^R : Kan^S ratio of 1:1 if the mutation does not affect gametophytic function (*see* **Note 12**). However, if the Kan^R : Kan^S is less than 1:1 ratio, it suggests that the mutation affects gametophytic function.

3.3. Phenotypic Analysis of Embryo Mutants by the Whole-Mount Clearing Method

1. Grow healthy wild-type and mutant plants under the same conditions.
2. Collect siliques at different developmental stages from the growing plants.
3. Open the siliques on a microscope slide along the septum line with a fine hypodermic needle under a binocular.

4. Remove the carpel wall from the siliques (*see* **Note 13**).
5. Add a drop of Herr's clearing solution on the ovules or seeds and cover them with a cover slip. Leave the slide at room temperature for a few minutes to several hours. Press the cover slip with a pencil or forcep to spread the seeds (optional).
6. Observe the cleared ovules under a Zeiss Axioskop II microscope equipped with DIC optics. Objectives 10×, 20×, and 40×.
7. Capture the images (*see* **Fig. 2**).

3.4. Cosegregation Analysis

In order to determine whether the mutant phenotype is due to the insertion of a T-DNA or *Ds* into a gene for embryo development, strict cosegregation of the phenotype and Kan resistance should be followed.

1. Grow healthy putative embryo mutant (heterozygous) in a growth chamber or a greenhouse.
2. Phenotypic analysis using the whole-mount clearing method (*see* **Subheading 3.3**) or reduced seed set (*see* **Subheading 3.1.2**). All the heterozygous Kan^R plants should segregate progeny with an embryo-defective phenotype.
3. Segregation ratio analysis of Kan selection marker (*see* **Subheading 3.1.1**). Self-progeny of heterozygous mutant should show a 2:1 of Kan^R : Kan^S segregation ratio for the embryo defective phenotype (*see* **Note 14**).

4. Notes

1. *fy60*, a recessive embryo-lethal mutant in our lab, is presented as an example for phenotypic analysis.
2. Unless stated otherwise, all solutions should be prepared in distilled water and should be stored at room temperature.
3. There should be enough seeds (>300) for segregation analysis.
4. Seed dormancy can be broken by keeping the seeds at 4 °C for 2–4 days before germination. This also results in synchronous seed germination.
5. The ratio of Kan^R : Kan^S seedlings can be used to identify lines that have a T-DNA or *Ds* insert disrupting a gene that is required for embryo development.
6. For a gametophyte-specific mutant, the Kan^R : Kan^S ratio is expected to approach 1:1 *(18,19)*. If the Kan^R : Kan^S ratio is >3, there is likely more than one T-DNA or *Ds* insertions into the genome.
7. Remember do not completely cover the plants.
8. Healthy plant is essential for seed set screen. Because reduced fertility can also be caused by other reasons such as excess watering and inappropriate growth conditions, each silique is expected to contain 50–65 ovules for healthy plants.
9. The optimal flowers selected for cross have to be at the stage when the white petals are just becoming visible. The papillae cells will not develop well if emasculated too early. Late emasculation might cause self-pollination. The forceps used should be cleaned carefully with 70% ethanol every time.

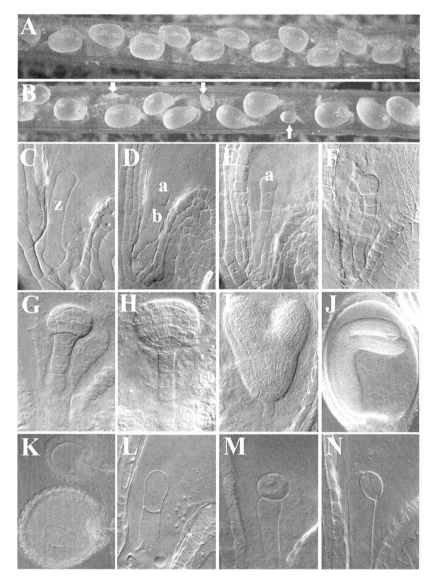

Fig. 2. Embryo development in *Arabidopsis* reviewed by whole-mount clearing. (**A**) A wild-type silique showing a full seed set. (**B**) A heterozygous *fy60* mutant silique showing a reduced seed set with approximately one-quarter of the embryos lethal (arrows). Wild-type embryos at the zygote (**C**), 1-cell (**D, E**), 4-cell (**F**), early globular (**G**), late globular (**H**), the heart (**I**), and the cotyledon (**J**) stages. (**K**) Whole-mounted, cleared seeds from siliques of heterozygous *fy60* plants. The same silique contains a mutant embryo arrested at the early globular stage (upper) compared with a normal embryo developed at the heart stage (lower). (**L–N**) Homozygous *fy60* embryos arrested at the 1-cell stage with aberrant cell division patterns. a, apical cell; b, basal cell; z, zygote.

10. Make sure do not hurt the stem when cutting off the siliques.

11. *Arabidopsis* is self-pollinated. It is important to remove the immature anthers from the female parent and introduce the pollen from another male parental line. In order to reduce the chance of cross-contamination, it is essential to keep mutant plants isolated from neighboring plants by covering a transparent plastic sleeve. Prepare several mutant plants to ensure that you have enough pistil that will mature and are intact.

12. This suggests that both the male and female gametophytes are functionally normal and the mutation can fully transmit to the next generation, indicating that the phenotype conferred by the mutant is attributable to a single recessive embryo-specific mutation.

13. Each silique checked is expected to contain 50–65 ovules.

14. All progeny from segregating sibling plants with full seed set are sensitive to Kan (wild type), but all the progeny from plants with embryo phenotype should show a 2:1 segregation for Kan resistance.

References

1. Page DR, Grossniklaus U. The art and design of genetic screens: *Arabidopsis thaliana*. Nat Rev Genet 2002;3:124–136.

2. Liu Y-G, Mitsukawa N, Oosumi T, Whittier RF. Efficient isolation and mapping of *Arabidopsis thaliana* T-DNA insert junctions by thermal asymmetric interlaced PCR. Plant J 1995;8:457–463.

3. Meinke DW, Sussex IM. Embryo-lethal mutants of *Arabidopsis thaliana*. Dev Biol 1979;72:50–61.

4. Meinke DW, Sussex IM. Isolation and characterization of six embryo-lethal mutants of *Arabidopsis thaliana*. Dev Biol 1979;72:62–72.

5. Franzmann LH, Yoon ES, Meinke DW. Saturating the genetic map of *Arabidopsis thaliana* with embryonic mutations. Plant J 1995;**7**:341–350.

6. Mcelver J, Tzafrir I, Aux G, Rogers R, Ashby C, Smith K, Thomas C, Schetter A, Zhou Q, Cushman MA, Tossberg J, Nickle T, Levin JZ, Law M, Meinke D, Patton D. Insertional mutagenesis of genes required for seed development in *Arabidopsis thaliana*. Genetics 2001;159:1751–1763.

7. Tzafrir I, Pena-Muralla R, Dickerman A, Berg M, Rogers R, Hutchens S, Sweeney TC, McElver J, Aux G, Patton D, Meinke D. Identification of genes required for embryo development in *Arabidopsis*. Plant Physiol 2004;135:1206–1220.

8. Ding Y-H, Liu N-Y, Tang Z-S, Liu J, Yang W-C. *Arabidopsis GLUTAMINE-RICH PROTEIN23* is essential for early embryogenesis and encodes a novel nuclear PPR motif protein that interacts with RNA polymerase II subunit III. Plant Cell 2006;18:815–830.

9. Springer PS, McCombie WR, Sundaresan V, Martienssen RA. Gene trap tagging of *PROLIFERA*, an essential *MCM2–3–5-like* gene in *Arabidopsis*. Science 1995;268:877–880.

10. Sundaresan V, Springer PS, Volpe T, Haward S, Jones JDG, Dean C, Ma H, Martienssen RA. Patterns of gene action in plant development revealed by enhancer trap and gene trap transposable elements. Genes Dev 1995;9:1797–1810.

11. Meinke DW. Perspective on genetic analysis of plant embryogeneisis. Plant Cell 1991;3:857–866.

12. Magnard JL, Lehouque G, Massonneau A, Frangne N, Heckel T, Gutierrez-Marcos JF, Perez P, Dumas C, Rogowsky PM. *ZmEBE* genes show a novel, continuous expression pattern in the central cell before fertilization and in specific domains of the resulting endosperm after fertilization. Plant Mol Biol 2003;53:821–836.

13. Herr JM Jr. A new clearing squash technique for the study of ovule development in angiosperms. Am J Bot 1971;58:785–790.

14. Vielle-Calzada JP, Baskar R, Grossniklaus U. Delayed activation of the paternal genome during seed development. Nature 2000;404:91–94.

15. Kyjovska Z, Repkova J, Relichova J. New embryo lethals in *Arabidopsis thaliana*: basic genetic and morphological study. Genetica 2003;119:317–325.

16. Murashige T, Skoog F. A revised medium for rapid growth and bioassays with tobacco tissue culture. Physiol Plant 1962;15:473–497.

17. Estelle MA, Somerville C. Auxin-resistant mutants of *Arabidopsis thaliana* with an altered morphology. Mol Gen Genet 1987;206:200–206.

18. Feldmann KA, Coury DA, Christianson ML. Exceptional segregation of a selectable marker (KanR) in *Arabidopsis* identifies genes important for gametophytic growth and development. Genetics 1997;147:1411–1422.

19. Shi DQ, Liu J, Xiang YH, Ye D, Sundaresan V, Yang WC. *SLOW WALKER1*, essential for gametogenesis in *Arabidopsis*, encodes a WD40 protein involved in 18S ribosomal RNA biogenesis. Plant Cell 2005;17:2340–2354.

9

Laser-Capture Microdissection to Study Global Transcriptional Changes During Plant Embryogenesis

Stuart A. Casson, Matthew W. B. Spencer, and Keith Lindsey

Summary

A key objective in the study of plant embryogenesis is to identify genes expressed in temporal and spatial patterns during development, in order to understand transcriptional control mechanisms regulating pattern formation, differentiation and morphogenesis. Mutagenic approaches have proved powerful to identify essential genes, but global, transcriptome-wide analysis of mRNA profiles in cells at different stages of differentiation would allow the identification of changes in the abundance of major classes of transcripts expressed from genes that are known to respond to regulatory signals, such as hormones. Particular classes of transcription factors or other genes might also be discovered to be associated with particular aspects of cell differentiation. This information would allow the construction of models to describe how signalling pathways might modulate transcriptional changes associated with cell differentiation. Previous limitations in tissue accessibility for RNA isolation have been overcome through the use of laser-capture microdissection, which allows cells from different embryonic tissues to be isolated, for RNA isolation, amplification and analysis by either polymerase chain reaction or DNA microarray techniques.

Key Words: *Arabidopsis thaliana*; embryogenesis; gene expression; laser-capture microdissection

1. Introduction

Laser-capture microdissection (LCM) is a relatively recent technological advance that allows the isolation of single cells or small clusters of cells for molecular or biochemical analysis, including the transcriptional profiling of cells. It was originally developed for the isolation of selected human cell populations from histological sections of complex, heterogeneous tissue. Subsequently,

From: *Methods in Molecular Biology, vol. 427: Plant Embryogenesis*
Edited by: M. F. Suárez and P. V. Bozhkov © Humana Press, Totowa, NJ

this technology has been applied successfully to a variety of animal systems *(1–3)*. Quick fixation or freezing of tissue samples for LCM minimizes any undesirable changes in gene expression that could occur during sample preparation *(4)*. LCM has been utilized extensively in pathology and cancer biology, providing insight through the isolation of cancer cells from complex tumour tissue *(5)* and is becoming of increasing importance in areas such as the study of neurodegenerative disease *(6)*. The objective of our own research has been to obtain tissue samples that would allow RNA extraction and amplification, for gene expression analyses in discrete regions of the developing *Arabidopsis* embryo *(7,8)*. This is an alternative approach to, for example, promoter trapping, which we have used extensively previously (e.g., *see* **refs. *9–13***). Different technologies now exist, which involve either cutting cells from sections using a UV laser (laser excision), such as afforded by the PALM® MicroLaser system (http://www.palm-mikrolaser.com/dsat/index.php), or pulling cells from sections (laser capture). We discuss here the latter system, which we have used extensively, with equipment developed by Arcturus (http://www.arctur.com).

To conduct LCM, a tissue section containing the cell type of interest is placed on a microscope stage. This is then covered with an isolation cap, which harbours a transparent thermoplastic polymer transfer film. An infrared laser beam is then targeted on the cells of interest, this activates the transfer film causing it to expand and impregnate the target cells, cementing the region of interest onto the cap. Due to the formation of a stronger adhesive force between the transfer cap and the tissue section than exists between the tissue section and the slide, the fused cells are torn out of the tissue section when the cap is lifted (*see* **Fig. 1**). This process can subsequently be repeated on new tissue sections to increase the number of cells acquired.

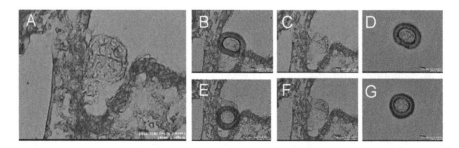

Fig. 1. Laser-capture microdissection of cryosections of a globular-stage *Arabidopsis* embryo. (**A**) Cryosectioned globular embryo. (**B**) Apical region after targeting with laser, (**C**) after removal of the cap and (**D**) apical cells captured on cap. (**E**) Targeting of the basal region, (**F**) after removal of the cap and (**G**) basal cells on the cap.

Until relatively recently animal tissue had been the exclusive subject for LCM, and protocols for fixation, sectioning and the extraction of RNA, DNA and proteins have been optimized for animal cells *(14)*. LCM is now beginning to be used in plant research, although structural and compositional differences between animal and plant cells have required the methodology to be adapted to account for the cell wall and vacuole. Protocols have now been developed in several plant species, including rice, maize and *Arabidopsis (7,8,15–20)*.

2. Materials
2.1. Cryosectioning

1. Tissue-Tek Optimal Cutting Temperature (OCT) embedding medium (RA Lamb).
2. Disposable base moulds, 15 × 15 × 5 mm (VWR).
3. Liquid nitrogen and liquid nitrogen-cooled isopentane (*see* **Note 1**).
4. Cryostat: for example, Leica CM3050S (Leica Microsystems).
5. Uncoated glass slides (VWR; *see* **Note 2**).
6. Ethanol solutions: 70% (v/v), 95% (v/v), 100% (v/v) stored at –22 °C (*see* **Note 3**).
7. Phosphate-buffered saline (PBS): Prepare 10× stock with 1.37 M NaCl, 27 mM KCl, 100 mM Na_2HPO_4, 18 mM KH_2PO_4 (adjust to pH 7.4 with HCl if necessary) and autoclave before storage at room temperature. Prepare working solution of 0.01% by dilution of one part with water.
8. Xylene. Store in fume cupboard.
9. Slide boxes containing silica gel (Sigma).

2.2. Laser-Capture Microdissection

1. PixCell II™ system (Arcturus).
2. CapSure™ HS LCM caps (Arcturus).
3. ExtracSure™ Sample Extraction Device (Arcturus).
4. Prep-strips™ (Arcturus).

2.3. RNA Extraction

1. RNA extraction kit for small tissue samples: Absolutely RNA™ Nanoprep kit (Stratagene; *see* **Note 4**).
2. β-Mercaptoethanol, 0.7 μL.
3. Ethanol, 70% (v/v).
4. Diethyl pyrocarbonate (DEPC)-treated water.

2.4. RNA Amplification

1. For cDNA synthesis and RNA amplification: MessageAmp™ aRNA kit (Ambion (Europe) Ltd).

3. Methods

LCM is a powerful tool and one that lends itself particularly to the isolation of embryonic regions, which previously could not be accessed by conventional dissection techniques. A key to its application in transcriptional profiling is in the successful amplification of the small amounts of RNA retrieved from captured tissues. The LCM method comprises four stages prior to gene expression analysis: tissue sectioning, cell capture, RNA extraction and RNA amplification. At this point gene expression analysis can be carried out by several possible techniques, such as reverse transcription-PCR, quantitative or semi-quantitative PCR, northern blot analysis or DNA microarray analysis. Tissue sectioning after fixing and embedding tissues gives excellent preservation of anatomical features, but detrimental effects on the quality of the RNA isolated have been demonstrated to arise from the use of such fixatives and the tissue manipulations involved in the embedding process *(4,14)*. In a study of tumour samples, OCT-embedded samples were shown to be comparable to fresh frozen tissue samples in terms of the gene expression profile observed from oligonucleotide microarray analysis *(21)*. Kerk et al. *(17)* investigated the RNA yield returned from tissue sections of differing thickness (3–10 μm) of radish (*Raphanus sativus)* cortical parenchyma and found little difference in this range. They standardized the section thickness used to 10 μm for mature tissue and 6 μm for developing tissue with comparatively smaller cells. The 8-μm section thickness routinely employed by use for analysis of embryonic tissues gives good results.

That the tissue histology was determined to be sufficient for this work is a consequence both of the small, cytoplasmically dense embryonic cells not being as susceptible to loss of cytological integrity as mature tissue and of limitations to the resolution of the LCM apparatus employed. The minimum beam setting on the Arcturus PixCell II system was only 7.5 μm, which ruled out the capture of single embryonic cells, and therefore the histology was only required to be sufficient to identify overall embryo morphology. Using a system capable of microdissecting a smaller area would potentially have necessitated a fixation step to fully exploit the capacity to target individual cells.

Maintaining the population distribution of transcripts is a major problem associated with the amplification of such small quantities of mRNA. Standard PCR methods are generally considered to produce a bias due to the preferential amplification of the smaller and more abundant transcripts in an mRNA population. Therefore, we use a linear T7 RNA polymerase-mediated amplification technique, as developed by Van Gelder et al. *(22)*, and which has previously been utilized to amplify RNA derived from animal tissue prior to microarray analysis. Therefore, following DNase treatment, RNA isolated from LCM cells was subjected to three rounds of amplification. This generated approximately 5–10 μg of amplified RNA (aRNA), with a size distribution of between 100 and 1000 nt *(7)*.

Nakazono et al. *(18)* compared T7-aRNA from laser microdissected material from maize coleoptiles with a comparable amount (40 ng) of non-aRNA using a two-colour cDNA microarray and demonstrated a highly linear relationship demonstrating reproducibility among samples. These studies only assessed changes to the expression profile after two rounds of amplification rather than the three used here. However, the results of Scheidl et al. *(23)* suggested that further rounds of amplification would produce no significant increase in variability. These authors suggested that the difference in expression profile observed between amplified and non-amplified samples results not from a bias resulting from experimental variability but from the global reduction in transcript length resulting from priming with random hexamers.

3.1. Sectioning

1. Remove siliques from plants at the appropriate stage of development, depending on the age of embryos required for analysis.
2. Dissect embryo sacs containing embryos from the siliques and embed in OCT-embedding medium in a base mould.
3. Freeze samples in liquid nitrogen-cooled isopentane prior to sectioning (*see* **Note 1**).
4. Cut 8-μm sections on a cryostat at −22 °C.
5. Collect sections on RNase-free uncoated glass slides and store in 70% (v/v) ethanol at −22 °C, in advance of further processing (*see* **Note 5**).
6. Sections are processed through the following ethanol dehydration series: 30 s in 70% (v/v) ethanol, 15 s in 0.01% PBS pH 7.4, 30 s in 70% (v/v) ethanol, 30 s in 95% (v/v) ethanol, 30 s in 100% ethanol, and finally two washes for 10 min each in xylene.
7. Slides are air-dried and can be stored dry in slide boxes containing silica gel (*see* **Note 6**).

3.2. Laser-Capture Microdissection

1. Slides are prepared for LCM using a prep-strip™ (Arcturus) to remove loosely adhered detritus.
2. Individual slides are placed into position on the LCM microscope stage.
3. A HS LCM cap is lowered onto the section using the placement arm.
4. The laser beam is set to 7.5 μm and focused with the 10× magnification lens, cells of interest are located and cell capture is performed with the laser beam set to either 7.5 or 15 μm. The power and duration of the laser beam was variable and dependent on beam size, but was typically 85 mW and 100 ms for a 7.5-μm beam (*see* **Note 7**).

5. Captured cells are then removed from the parent tissue section by rapidly lifting the LCM cap using the placement arm. Typically, around 15 sections were processed per LCM cap, and non-specific material was removed from the surface of a LCM cap using a Post-It note (*see* **Note 8**).

6. An ExtracSure™ Sample Extraction Device (Arcturus) is then attached to the LCM cap in preparation for RNA extraction.

3.3. RNA Extraction

1. RNA from LCM cells is extracted using the Absolutely RNA™ Nanoprep kit (Stratagene) according to the manufacturer's instruction with slight modifications. In brief, apply 100 μL of lysis buffer and 0.7 μL β-mercaptoethanol applied to the captured cells on the cap via the ExtracSure™ Sample Extraction Device, which is then connected to a 0.5-mL microcentrifuge tube.

2. Vortex to mix the sample and incubate it at 60 °C for 5 min (*see* **Note 9**).

3. Collect the lysis buffer and mix with an equal volume (~100 μL) of 70% (v/v) ethanol.

4. Apply the extract to an RNA-binding column and centrifuge according to the manufacturer's instruction.

5. Treat with DNase to remove contaminating gDNA.

6. Wash the RNA extract and elute in 2 × 20 μL of DEPC-treated water (*see* **Note 10**).

7. Use vacuum concentration to bring the volume of RNA down to 11 μL, ready for amplification (*see* **Note 11**).

3.4. RNA Amplification

1. cDNA synthesis and RNA amplification are performed using specialized kits for working with very small samples, such as the MessageAmp™ aRNA kit (Ambion (Europe) Ltd). Reactions are carried out in a thermal cycler.

2. First round cDNA synthesis is primed with a T7 oligo (dT) primer. Add 1 μL T7 oligo (dT) primer and 11 μL RNA and heat to 70 °C for 10 min.

3. At 42 °C, add 2 μL 10x first strand buffer, 1 μL ribonuclease inhibitor, 4 μL dNTP mix and 1 μL reverse transcriptase and incubate for 2 h.

4. Second strand synthesis is performed by the addition of 63 μL nuclease-free water, 10 μL 10x second strand buffer, 4 μL dNTP mix, 2 μL DNA polymerase and 1 μL RNase H. Incubate for 2 h at 16 °C.

5. Purify cDNA using the binding columns provided according to the manufacturer's instructions and vacuum concentrate to 16 μL.

6. For in vitro transcription, mix 8 μL of the cDNA with 8 μL dNTP mix, 2 μL 10x reaction buffer and 2 μL T7 enzyme mix. Incubate at 37 °C for 16–24 h and then treat by DNase I digestion (*see* **Note 12**).

7. aRNA is purified and vacuum concentrated to 10 μL.

8. Subsequent rounds of cDNA synthesis are performed as follows. Add 2 μL of random primers to 10 μL of aRNA and incubate the reaction at 70 °C for 10

min. At 42 °C, add the following: 2 μL 10× first strand buffer, 1 μL ribonuclease inhibitor, 4 μL dNTP mix and 1 μL reverse transcriptase. Incubate the reaction for 2 h (*see* **Note 13**).

9. Add 1 μL of RNase H and incubate for 30 min at 37 °C.

10. Prime second strand synthesis with 5 μL T7 oligo (dT) primer and incubate for 10 min at 70 °C.

11. Then add the remaining reagents at room temperature: 58 μL nuclease-free water, 10 μL 10× second strand buffer, 4 μL dNTP mix, 2 μL DNA polymerase. Incubate the reaction at 16 °C for 2 h. cDNA is then purified and in vitro transcription performed as described in **steps 6** and **7**.

12. Typically three rounds of RNA amplification are required to produce microgram quantities of aRNA (*see* **Note 14**).

4. Notes

1. Liquid nitrogen-cooled isopentane provides more efficient heat transfer, therefore reducing ice crystal formation, compared to freezing in liquid nitrogen alone. To prepare, in a fume cupboard pour approximately 100 mL of isopentane into a 150-mL Pyrex beaker. Using a clamp and retort stand, suspend the beaker of isopentane into a dewar containing liquid nitrogen and allow to cool for 10 min. The isopentane will eventually freeze completely, but this can be avoided by periodically removing it from the liquid nitrogen and stirring with a glass rod. Using forceps, plunge the mould containing the OCT-embedded sample to the bottom of the liquid nitrogen-cooled isopentane for approximately 30 s (the OCT turns white when frozen). The frozen sample can then be quickly placed in liquid nitrogen whilst more moulds are prepared. Sample blocks can be used immediately or stored at –80 °C until required. Sample blocks should be placed in the cryostat and allowed approximately 15 min to acclimatize to –22 °C prior to trimming and sectioning.

2. We would clean slides by washing in acetone in slide containers followed by baking at 200 °C. Uncoated slides were generally found to give the best balance between sample adhesion and the ability to remove tissue by LCM. If sample adhesion becomes an issue, Polylysine slides (VWR) can be used.

3. For successful LCM, samples must be completely dehydrated; therefore, 100% ethanol should be prepared in advance using molecular sieves, 1/8 inch pellets (Sigma).

4. Several manufacturers now produce RNA extraction kits specifically aimed for application with LCM tissue. Due to the small quantities of RNA isolated from LCM tissues, we advise the use of one of these kits as opposed to standard RNA precipitative methods unless a precipitation aid, such as carrier yeast tRNA, is added.

5. Several sections can be collected per slide, especially if you can cut ribbons. Take a room temperature slide and press it down firmly onto the cut sections on the back plate of the cryostat. Lift the slide and turn it over onto the back plate

for approximately 15 s to re-freeze the sections. The slide can then be placed in 70% ethanol. It is best to have this within the cryostat itself. At this point, sections can be quickly checked under a microscope prior to further processing, or this can wait until after processing.

6. Complete dehydration of sections is required for successful LCM so it is important that, following the xylene treatment, the sections do not absorb any atmospheric moisture.

7. For laser focusing, you must focus the microscope on the black ring of the LCM HS cap and not the specimen. You will need to reduce the illumination such that the beam appears as a white spot. Focus until you have an intense, tight spot with a faint halo. If you decide to change the spot size to 15 μm, the laser will have to be refocused (always on 10× magnification). When focused, try some practice 'shots' on a region of the cap devoid of tissue. You are looking for the membrane to melt cleanly without any scorching, which occurs if the power or duration is too high. Once you are satisfied that the settings are correct, proceed to capturing the tissue. One cap can be used to capture cells from several sections. If the laser does not cut after a capture, make sure that the cap is sitting correctly in the placement arm.

8. On occasion, tissue capture will result in unwanted cells adhering to the cap. These cells are only loosely stuck on the cap surface and are not as securely tethered as the LCM-targeted cells. The easiest way to remove these cells is to use the adhesive on a Post-It note. Gently place the cap on the adhesive surface and slowly peel it off. Examine the cap surface to determine if unwanted cells have been removed and repeat if required. This approach is not detrimental to the quality of the RNA, though it is not recommended for protein extraction.

9. We found that a brief heating step improved the extraction process. Remember that extraction buffer must always be in contact with the cap surface.

10. A larger volume of 20 μL is suggested as we observed insufficient wetting of the column with the manufacturer's recommended volume.

11. As with **Note 4**, we would advise avoiding a precipitation step at this point due to the small quantities of RNA involved. If you do not have access to a vacuum concentrator, then add a suitable amount of carrier tRNA or glycogen to aid precipitation.

12. Only half the cDNA is being used as a template for in vitro transcription, with the remaining half stored at −80 °C. If this archive is not required, at step 5 concentrate the cDNA to 8 μL and use all of it in step 6. The long incubation time improves aRNA yield; however, the reaction is sensitive to evaporation. We suggest overlaying the in vitro transcription reaction with RNase-free mineral oil (Sigma).

13. It is important to remember that after the first round of cDNA synthesis and in vitro transcription, the orientation of the aRNA is reversed, and therefore for first strand cDNA synthesis, random primers are used with the T7 oligo (dT) priming the second strand synthesis.

14. You can monitor an aliquot of your aRNA after each round of amplification by using standard spectroscopy, a NanoDrop ND-1000 Spectrophotometer, an Agilent Bioanalyser, or gel electrophoresis. This will allow you to judge when you have enough aRNA for your required purpose.

Acknowledgments

The authors would like to thank BBSRC for funding their work on LCM and plant embryogenesis.

References

1. Emmert-Buck MR, Bonner RF, Smith PD, Chuaqui RF, Zhuang ZP, Goldstein SR, Weiss RA, Liotta LA. Laser capture microdissection. Science 1996;274:998–1001.
2. Bonner RF, Emmert-Buck MR, Cole K, Pohida T, Chuaqui RF, Goldstein SR, Liotta LA. Cell sampling – laser capture microdissection: molecular analysis of tissue. Science 1997;278:1481–1483.
3. Simone NL, Bonner RF, Gillespie JW, Emmert-Buck MR, Liotta LA. Laser-capture microdissection: opening the microscopic frontier to molecular analysis. Trends Genet 1998;14:272–276.
4. Gillespie JW, Best CJM, Bischel VE, Cole KA, Greenhut SF, Hewitt SM, Ahram M, Gathright YB, Merino MJ, Strausberg RL, Epstein JI, Hamilton SR, Gannot G, Baibakova GV, Calvert VS, Flaig MJ, Chuaqui RF, Herring JC, Pfeifer J, Petricoin EF, Linehan WM, Duray PH, Bova GS, Emmert-Buck MR. Evaluation of non-formalin tissue fixation for molecular profiling studies. Am J Pathol 2002;160: 449–457.
5. Gillespie JW, Ahram M, Best CJ, Swalwell JI, Krizman DB, Petricoin EF, Liotta LA, Emmert-Buck MR. The role of tissue microdissection in cancer research. Cancer J 2001;7:32–39.
6. Standaert DG. Applications of laser capture microdissection in the study of neurodegenerative disease. Arch Neurol 2005;62:203–205.
7. Casson S, Spencer M, Walker K, Lindsey K. Laser capture microdissection for the analysis of gene expression during embryogenesis of *Arabidopsis*. Plant J 2005;42:111–123.
8. Spencer MWB, Casson SA, Lindsey K. Transcriptional profiling of the *Arabidopsis thaliana* embryo. Plant Physiol 2007;143:924–940.
9. Topping JF, Wei W, Lindsey K. Functional tagging of regulatory elements in the plant genome. Development 1991;112:1009–1019.
10. Topping JF, Agyeman F, Henricot B Lindsey K. Identification of molecular markers of embryogenesis in *Arabidopsis thaliana* by promoter trapping. Plant J 1994;5:895–903.
11. Topping JF, Lindsey K. Promoter trap markers differentiate structural and positional components of polar development in *Arabidopsis*. Plant Cell 1997;9:1713–1725.

12. Casson SA, Chilley PM, Topping JF, Evans IM, Souter MA, Lindsey K. The *POLARIS* gene of Arabidopsis encodes a predicted peptide required for correct root growth and leaf vascular patterning. Plant Cell 2002;14:1705–1721.

13. Farrar K, Evans IM, Topping JF, Souter MA, Nielsen JE, Lindsey K. *EXORDIUM* – a gene expressed in proliferating cells and with a role in meristem function, identified by promoter trapping in *Arabidopsis*. Plant J 2003;33:61–73.

14. Goldsworthy SM, Stockton PS, Trempus CS, Foley JF, Maronpot RR. Effects of fixation on RNA extraction and amplification from laser capture microdissected tissue. Mol Carcinog 1999;25:86–91.

15. Matsunaga S, Schütze K, Donnison IS, Grant SR, Kuroiwa T, Kawano S. Single pollen typing combined with laser-mediated manipulation. Plant J 1999;20: 371–378.

16. Asano T, Masumura T, Kusano H, Kikuchi S, Shimada H, Kadowaki K-I. Construction of a specialized cDNA library from plant cells isolated by laser capture microdissection: toward comprehensive analysis of the genes expressed in the rice phloem. Plant J 2002;32:401–408.

17. Kerk NM, Ceserani T, Tausta SL, Sussex IM, Nelson TM. Laser capture microdissection of cells from plant tissues. Plant Physiol 2003;132:27–35.

18. Nakazono M, Qui F, Borsuk LA, Schnable PA. Laser-capture microdissection, a tool for the global analysis of gene expression in specific plant cell types: identification of genes expressed differentially in epidermal cells or vascular tissues of maize. Plant Cell 2003;15:583–596.

19. Liu X, Wang H, Li Y, Tang Y, Liu Y, Hu X, Jia P, Ying K, Feng Q, Guan J, Jin C, Zhang L, Lou L, Zhou Z, Han B. Preparation of single rice chromosome for construction of a DNA library using a laser microbeam trap. J Biotechnol 2004;109:217–226.

20. Inada N, Wildermuth MC. Novel tissue preparation method and cell-specific marker for laser microdissection of *Arabidopsis* mature leaf. Planta 2005;221:9–16.

21. Sanchez-Carbayo M, Saint F, Lozano JJ, Viale A, Cordon-Cardo C. Comparison of gene expression profiles in laser-microdissected, nonembedded, and OCT-embedded tumour samples by oligonucleotide microarray analysis. Clin Chem 2003;49:2096–2100.

22. Van Gelder RN, Von Zastrow ME, Yool A, Dement WC, Barchas JD, Eberwine JH. Amplified RNA synthesized from limited quantities of heterogeneous cDNA. Proc. Natl. Acad. Sci. U. S. A. 1990;87:1663–1667.

23. Scheidl SJ, Nilsson S, Kalen M, Hellstrom M, Takemoto M, Hakansson J, Lindahl P. mRNA expression profiling of laser microbeam microdissected cells from slender embryonic structures. Am J Pathol 2002;160:801–813.

10

Promoter Trapping System to Study Embryogenesis

Robert Blanvillain and Patrick Gallois

Summary

Promoter trapping is a particular gene trap strategy that represents a valuable tool for the discovery of specific cell-type markers. The principle is to generate a collection of transgenic lines with random insertions of a promoter-less reporter gene and to screen for specific reporter activity in the domain of interest. The use of β-glucuronidase (*GUS*) as a reporter gene provides a simple and sensitive assay that allows identification of very restricted expression patterns and makes the promoter trap appropriate to study embryogenesis. Plant embryogenesis starts at the fertilization of the egg cell encapsulated in the maternal tissue and leads to the establishment of a new organism capable of an autonomous life. Uncovering genes specifically expressed in sub-domain of the embryo during its development represents a major technical challenge due, in part, to size and accessibility limitations. Promoter trapping approaches have been successfully used to overcome these problems. The trapped activity represents thereafter a useful genetic marker of the uncovered cell type, which is expected to reveal the properties of a specific promoter shedding light on a new gene function. In this chapter, protocols for examining and documenting *GUS* reporter gene activities in the embryo are described. Methods for the amplification of sequences flanking insertions and subsequent molecular and genetic characterization are provided.

Key Words: *Arabidopsis*; embryogenesis; promoter trap; GUS assay; TAIL-PCR; seed clarification.

1. Introduction

Gene and enhancer trap insertions allow the identification of genes based on the expression pattern of a reporter gene. The principle of the strategy relies on the creation of random genomic insertions of a reporter gene, the expression of which can be easily visualized. In a subset of insertions, regulatory elements of a nearby

From: *Methods in Molecular Biology, vol. 427: Plant Embryogenesis*
Edited by: M. F. Suárez and P. V. Bozhkov © Humana Press, Totowa, NJ

Fig. 1. Schematic drawing showing the T-DNA used in the pDeltaGUS promoter trap system *(8,9)*. The *NptII* gene, conferring kanamycin resistance, and the *GUS* gene (*uidA*) are shown, with arrows indicating the direction of transcription. It is important to avoid the presence of 35S enhancers in the T-DNA construct. 35S enhancers are known to influence distant promoters and are likely to modify the expression pattern of the trapped promoter.

chromosomal gene transcriptionally activate the reporter gene. It is expected that the observed activity would accurately report the expression pattern of the trapped gene, and screens are specifically designed to allow the identification of temporally and spatially regulated marker genes. This approach has been successfully used in a number of different organisms, including bacteria, *Drosophila*, mice, and plants *(1–4)*. In plants, most trapping systems have been developed to take advantage of transposons or T-DNA as the insertion agent and *GUS* as the reporter gene of choice, reviewed in **ref.** *(5)*. The green fluorescent protein has also been utilized as reporter gene, for example, *see* **ref.** *(6)*. The trap construct can be designed as a promoter trap, a gene trap or an exon trap, reviewed in **ref.** *(7)*. Our group and other groups have successfully used promoter trap or gene trap approaches to study embryogenesis and seed development *(8–12)*. Although there are some instances where a trap construct is reported to have identified a cryptic promoter rather than a gene *(13,14)*, there is plenty of evidence in the literature that bona fide genes can be tagged *(12,15)*.

We describe here methods for GUS staining, molecular characterization, and genetic analysis of T-DNA insertion revealing promoter activity during embryogenesis based on a promoter-trap system that we used in the model plant *Arabidopsis thaliana* to study embryogenesis (*see* **Fig. 1**) *(8)*. These techniques could also be used to screen our collection and many generated collections available at the Arabidopsis Biological Resource Center. We have focused on the most common techniques used to identify reporter gene expression patterns during *Arabidopsis* embryogenesis using GUS staining, which are complementary to the published *Arabidopsis* protocols manual *(16)*.

2. Materials

2.1. Screening for GUS Expression Patterns

2.1.1. Silique Dissection

1. Tweezers Dumont no. 5.
2. Dissecting Scissors (F.S.T, cat. no. 5018-10).

3. Needles Sub-Q 26G5/6.
4. 3MM filter discs.
5. Dissecting scope.

2.1.2. GUS Staining

1. Acetone, 90%.
2. 1 M sodium phosphate buffer, pH 7.0: 57.7 mL 1 M Na_2HPO_4, 42.3 mL 1 M NaH_2PO_4 (autoclaved).
3. GUS staining solution: 100 mM sodium phosphate buffer, pH 7.0 (from 1 M stock), 10 mM ethylene diamine tetraacetic acid (EDTA), 0.1% Triton® X-100, 1 mg/mL 5-bromo-4-chloro-3-inolyl-β-D-glucuronic acid, cyclohexylammonium salt (X-Gluc), 2 mM potassium ferricyanide, 2 mM potassium ferrocyanide. X-Gluc dissolves in dimethylformamide at 100 µg/mL (*see* **Note 1**).
4. 48-well tissue culture plates.
5. Vacuum desiccator.
6. Ethanol, 70%.
7. Glycerol, 50% (v/v).
8. Microscope equipped with Nomarski optics for differential interference contrast (DIC) microscopy.

2.1.3. Seed Coat Clarification

1. Hoyer's medium: 8 g Chloral hydrate, 1 mL 100% glycerol, 2 mL distilled water (*see* **Note 2**).
2. Glass slides.
3. Cover glasses, no. 1.5, or frame seals 125 µL (cat. no. 0578, AB).

2.1.4. Sectioning of GUS-Stained Siliques

1. 100 mM sodium phosphate buffer, pH 7.0 (from 1 M stock, *see* **Subheading 2.1.1**).
2. Paraformaldehyde fixative: 4% paraformaldehyde in 100 mM sodium phosphate buffer, pH 7.0 (*see* **Note 3**).
3. Phosphate-buffered saline (PBS): 137 mM NaCl, 2.7 mM KCl, 10 mM Na_2HPO_4, 2 mM KH_2PO_4, adjust pH to 7.4 with HCl. Can be made as a 10× stock and be autoclaved.
4. Ethanol: 100%, 95%, 80%, 70%, 50%, and 30%.
5. Eosin, dilute at 0.1% (w/v) in 100% ethanol.
6. Histoclear clearing agent.
7. Paraplast Plus chips (cat. no. 23–021400, Fisher Scientific).
8. 60 °C Oven.
9. Glass scintillation vials.
10. Rotary microtome.

11. Coated microscope slides (cat. no. 12–550–15, Fisherbrand Superfrost Plus, Fisher Scientific).
12. Variable temperature hot plate.
13. Mount-Quick aqueous mounting media (cat. no. 18002, EM Science).

2.2. Amplification of DNA Sequences Flanking Insertion

2.2.1. DNA Isolation

1. Kontes disposable pellet pestle (Fisher Scientific; cat. no. K749521–1590).
2. Cetyl trimethyl ammonium bromide (CTAB) extraction buffer (2× CTAB): 2% (w/v) CTAB 100 mM Tris–HCl, pH 8.0, 20 mM EDTA, pH 8.0, 1.4 M NaCl, 2% (w/v) PVP 40.
3. 2-β-mercaptoethanol.
4. Chloroform : isoamylalcohol, 24:1, no phenol.
5. Isopropanol.
6. Sodium acetate, 3 M, pH 5.2.
6. TE: 10 mM Tris–HCl, pH 8.0, 1 mM EDTA.
7. 100% and 70% Ethanol.

2.2.2. Thermal Asymmetric Interlaced Polymerase Chain Reaction

1. *T-DNA*-specific primers, 10 μM stocks (*see* **Note 4**). **GPTV-L1** (5′-TGCTTTACGGCACCTCGAC-3′); **GPTV-L2** (5′-TGGTTCACGTAGTGGGC CATCG-3′); **GPTV-L3** (5′-GCGTGGACCGCTTGCAACT-3′).
2. Arbitrary degenerate (AD) 128-fold primers for thermal asymmetric interlaced polymerase chain reaction (TAIL-PCR), 12 μM stocks, where S is G or C, W is A or T, N is A, T, C, or G. **L-AD2** (5′-NGTCGASWGANANGAA-3′) *(17)*; **LW-AD1** (5′-TGWGNASANCASAGA-3′) *(18)*; **LW-AD2** (5′-AGWG NAGWANCAWAGG-3′) *(18)*.
3. *Taq* DNA polymerase, 5 U/μL.
4. PCR buffer, 10× (supplied with *Taq* DNA polymerase): 500 mM KCl, 15 mM MgCl₂, 100 mM Tris–HCl, pH 9.0.
5. Deoxyribonucleotide (dNTP) mixture containing 5 mM each of dATP, dTTP, dGTP, and dCTP.

2.3. Sequencing of TAIL-PCR Products

1. Shrimp Alkaline Phosphatase (SAP) (Promega).
2. *E. coli* Exonuclease I (ExoI) (Promega).
3. GPTV-L3 primer, 10 μM (*see* **Subheading 2.2.2**).
4. BigDye® Terminator v1.1 Cycle Sequencing Kit (Applied Biosystems).

2.4. Sequence Analysis

Access to the Internet to retrieve sequence identity using Basic Local Alignment Search Tool (BLAST) and the *Arabidopsis* information resources (TAIR) at http://www.arabidopsis.org/.

3. Methods

3.1. Screening for GUS Expression Patterns During Embryogenesis

Histochemical detection of GUS activity in most *Arabidopsis* tissues is very straightforward. In embryos, however, GUS staining requires some manipulations of the seedpod.

3.1.1. Silique Dissection

1. In order to cover the embryogenesis stages of interest, grow the plant on soil to allow the inflorescence to fully develop.
2. Tag the open flowers every day with a different color thread that allows a later simultaneous harvesting of consecutive stages.
3. Harvest siliques when appropriate, cut the stigma with dissecting scissors, and transversally cut with a needle along the midrib to expose the fertilized ovules without damaging the funiculus (*see* **Fig. 2**). The ovules should remain attached to the placenta (*see* **Note 5**).
4. Place directly the open silique into cold acetone 90% for 30 min (acetone should be stored at −20 °C prior to use).
5. Wash three times at least 30 min in abundant 1× phosphate buffer pH 7.0 (*see* **Note 6**).

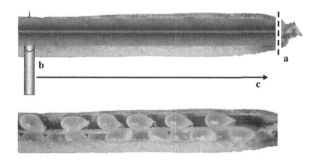

Fig. 2. Procedure used to open a silique. (a) Remove the stigmatic end of the silique to ease the passage of the needle. (b) While holding the silique at the pedicel with forceps, insert the needle below the midrib but above the seeds and (c) cut along the midrib and remove the cut stripe; bottom panel is the resulting open silique.

3.1.2. GUS Staining

1. Incubate the siliques in GUS stain solution, making sure they are entirely submerged. Because the promoter-trap insertion will be segregating in T2 seeds, it makes sense to screen the T1 seeds growing on the first transformed generation. To minimize expenses, use just enough stain solution to cover the tissue. Siliques can be stained in 24- or 48-well tissue culture plates.

2. Place the tissue culture plate in a vacuum desiccator and draw a vacuum for 15 min.

3. Release the vacuum, seal the plates with parafilm or tape, and wrap them in aluminum foil to exclude light.

4. Incubate the staining reaction at 37 °C for 16–24 h. For initial screens, we recommend incubating for at least 24 h to allow detection of GUS activity in lines that have low levels of *GUS* expression. The long incubation, however, may result in overstaining and decreased specificity in lines showing high levels of *GUS* expression. Lines of interest should be examined in a secondary screen using suitable incubation times. In each experiment, include a control line with a characterized *GUS* expression pattern.

5. Remove the GUS stain solution and replace it with 70% ethanol to remove chlorophyll from the tissue.

6. Place stained tissues in 70% ethanol in a Petri dish for manipulation. Extract the seeds from the silique using a needle and examine *GUS* expression patterns. This is often the most time-consuming step in a large-scale screen. Be sure to allow adequate time for careful examination. It is convenient to perform the initial screen using a stereomicroscope. Seeds can also be mounted on a glass slide in 50% glycerol and viewed using a compound microscope with DIC optics. In most cases, the tissue can be transferred directly from 70% ethanol to 50% glycerol, skipping re-hydration, without a substantial loss of tissue integrity.

3.1.3. Seed Coat Clarification

It can be challenging to obtain good images of GUS-stained embryos, especially in the case of very faint expression patterns in mature stages. It is almost always necessary to go through seed coat clarification.

1. Incubate in 70% ethanol at room temperature, changing the ethanol when it turns green, as many times as necessary (as a guide: two or three times for 60 min). This step is largely eased by the acetone pre-treatment; it is important to adequately clear chlorophyll from the tissue in order to visualize faint *GUS* expression patterns.

2. Incubate in Hoyer's medium for as long as necessary to clarify the seed coat. This step can take from 1 h up to 2 days and can be directly done on the slide using a 125-μL frame seal (*see* **Note 7**).

3. View the tissue using a microscope with DIC optics.

3.1.4. Sectioning GUS-Stained Siliques

Higher resolution visualization of *GUS* expression patterns can be obtained using tissue sections.

1. Wash the GUS-stained siliques in 100 mM sodium phosphate buffer, pH 7.0, 1 h to overnight.
2. Replace the phosphate buffer with 4% paraformaldehyde fixative and vacuum infiltrate the tissue on ice. Allow tissue to fix for 60 min.
3. Wash with ice-cold PBS, 3×, for 10 min.
4. Dehydrate the tissue through an ethanol series (30%, 50%, 70%, 80%, 95%, and 100%), 60 min each step at 4 °C.
5. Replace the ethanol with 0.1% (w/v) eosin in 100% ethanol and leave at 4 °C overnight.
6. Perform an additional incubation in 100% ethanol for 60 min at room temperature to be sure that no water remains.
7. Replace the ethanol with Histoclear using a series of Histoclear/ethanol mixtures at room temperature: 25% Histoclear/75% ethanol for 60 min, 50% Histoclear/50% ethanol for 60 min, 75% Histoclear/25% ethanol for 60 min, 100% Histoclear 3× for 60 min.
8. Replace Histoclear and add 25% final vol. of Paraplast wax chips and incubate at 42 °C until the chips have melted.
9. Add another 25% vol. of Paraplast chips and move to 60 °C.
10. Replace wax/Histoclear mixture with freshly melted wax.
11. Replace with freshly melted wax twice a day for 3 days. The Paraplast temperature should not exceed 60 °C. Work quickly when changing wax, as it will solidify rapidly.
12. Pour tissue into molds on a variable temperature hot bench and position using hot needles or tweezers.
13. Move gradually the mold to cooler positions on the bench and orient the siliques.
14. When the wax hardens, cut blocks that contain the tissue, mount, trim, and section using a rotary microtome. Best results are obtained when the blade is parallel to the axis of the septum. For embryos, you may want to cut fairly thick sections (>10 µm).
16. Float the sections on a slide in water at 42 °C on a slide warmer.
17. When the sections have expanded, remove the water and dry overnight on the slide warmer.
18. Deparaffinize them by soaking in Histoclear, 2×, for 10 min.
19. Rehydrate the tissue through a series of ethanol steps: 100%, 95%, 80%, 70%, 50%, and 30%, 10 min each, then transfer to water.
20. Mount under a cover glass in aqueous mounting media (*see* **Note 8**).
21. Allow the mounting media to dry overnight and observe with the microscope using DIC optics.

3.2. Amplification of DNA Sequences Flanking T-DNA Insertion

3.2.1. DNA Isolation

Genomic DNA from wild-type and the transgenic lines of interest can be isolated from seedlings or a few inflorescences.

1. Harvest tissue, pooling at least four different transgenic plants if the T-DNA segregates.
2. Grind the tissue in 250 µL of 2× CTAB buffer in a 1.5-mL microcentrifuge tube using a disposable pellet pestle.
3. Add 250 µL of 2× CTAB buffer and 2 µL of 2-β-mercaptoethanol.
4. Mix and incubate the extracts at 65 °C for 30 min.
5. Add 300 µL of chloroform : isoamylalcohol (24:1) in a chemical fume hood and vortex for 3 min. Wear gloves.
6. Spin in a microcentrifuge at 10,000 g for 8 min.
7. Transfer the top (aqueous) phase to a new microcentrifuge tube and add 50 µL (1/10 volume) of 3 M sodium acetate, pH 5.2, followed by 300 µL of 100% isopropanol to precipitate nucleic acids. Mix.
8. Incubate at room temperature for 15 min.
9. Spin for 8 min at 10,000 g to pellet the DNA.
11. Remove the supernatant and rinse the pellet with 500 µL of 70% ethanol.
12. Centrifuge 10,000 g for 5 min in a microcentrifuge.
13. Remove the supernatant and allow pellet to air-dry on the bench for about 30 min.
14. Resuspend the pellet in 20 µL sterile milliQ water or TE.
15. Quantify the DNA by electrophoresis of 2 µL of each DNA sample on a 0.8% agarose gel. To estimate the DNA concentration, compare the band intensity to that of DNA fragments of known concentration using a commercial DNA ladder. Expect a single faint narrow band of high molecular weight. It is also advised to examine the DNA quality using an analytical PCR with control primers.

3.2.2. TAIL-PCR

Amplification of genomic sequences flanking the T-DNA can be performed by TAIL-PCR. Three successive PCRs are performed using AD primers and nested primers specific to the right border of the T-DNA. The primers GPTV-L1, GPTV-L2, and GPTV-L3 are nested primers for the right border end of the T-DNA.

1. Primary TAIL-PCR (TAIL-1). Add 5 µL of DNA to a set of three PCRs of 20 µL each which contain 0.4 µL of 10 µM GPTV-L1, 5 µL of one of the three 12 µM degenerate primers, 0.8 unit of Taq-polymerase, 0.8 µL of 5 mM dNTPs, 2 µL of 10× Buffer, 6 µL of sterile milliQ water. The TAIL-1 PCR program includes the following steps: 93 °C for 1 min; 95 °C for 1 min; six cycles (94 °C for 30 s, 62 °C for 1 min, and 72 °C for 2.5 min); 94 °C for 30 sec, 25 °C for 3 min, 72 °C

for 2.5 min; 15 cycles (94 °C for 30 s, 68 °C for 1 min, 72 °C for 2.5 min, 94 °C for 30 s, 68 °C for 1 min, 72 °C for 2.5 min, 94 °C for 30 s, 44 °C for 1 min, and 72 °C for 2.5 min); 72 °C for 2.5 min.

2. Secondary TAIL-PCR (TAIL-2). After the first amplification step, dilute 1 μL of each PCR with 199 μL of sterile milliQ water and add an aliquot of 4 μL from each 200-fold diluted sample to a second PCR of 20 μL which contained 0.4 μL of 10 μM GPTV-L2, 5 μL of the 12 μM degenerate primer used in TAIL-1, 0.6 unit of Taq-polymerase, 0.8 μL of 5mM dNTPs, 2 μL of 10× Buffer, 7.2 μL of sterile water. The TAIL-2 PCR program is 13 cycles (94 °C for 30 s, 64 °C for 1 min, 72 °C for 2.5 min, 94 °C for 30 s, 64 °C for 1 min, 72 °C for 2.5 min, 94 °C for 30 s, 44 °C for 1 min, and 72 °C for 2.5 min) and 72 °C for 5 min.

3. Tertiary TAIL-PCR (TAIL-3). After the TAIL-2 amplification step, dilute 1 μL of each PCR with 99 μL of sterile milliQ water and add 5 μL from each 100-fold diluted sample to a third PCR of 50 μL which contain 1 μL of 10 μM GPTV-L3, 12.5 μL of the 12 μM degenerate primer used in original TAIL-1, 1.5 unit of Taq-polymerase, 2 μL of 5mM dNTPs, 5 μL of 10× Buffer, 22.7 μL of sterile milliQ water. The TAIL-3 PCR program is 20 cycles (94 °C for 10 s, 44 °C for 1 min, and 72 °C for 2.5 min) and 72 °C for 5 min.

4. Analyze the result of each reaction by running 5–10 μL of the TAIL-2 and TAIL-3 PCR products in adjacent lanes using a 1.5% agarose gel. Successful amplification will result in the presence of visible products in both reactions, with a characteristic size shift between the secondary and tertiary products due to the use of nested T-DNA primers (*see* **Note 9**).

3.3. Sequencing of TAIL-PCR Products

When a TAIL-3 reaction shows a single major PCR product, it can be purified for direct sequencing using enzymatic degradation of primers and dNTPs.

1. Add 2 μL of an enzymatic cocktail containing 0.5 units of the SAP and 2 units ExoI to 15.5 μL of the TAIL-3 PCR products.

2. Incubate 1 h at 37 °C. Heat-inactivate the enzymes at 70 °C for 20 min. It may be necessary to repeat the tertiary TAIL-PCRs to obtain a sufficient amount of PCR product for sequencing (*see* **Note 10**).

3. For the sequencing reaction, add 0.6 μL of 10 μM GPTV-L3 and 4 μL of BigDye® to the 15.5 μL of the tertiary TAIL-PCR products purified with the SAP-ExoI enzymatic cocktail. The sequencing reaction program includes the following steps: 95 °C for 5 min; 35 cycles (95 °C for 30 s, 55 °C for 30 s, and 60 °C for 3 min).

4. Add 4 μL of 3 M sodium acetate pH 5.2, 80 μL of 100% ethanol, 20 μL of water and 1 μL of glycol blue to stain the DNA pellet to 20 μL of the sequencing reaction. Incubate 10 min at room temperature and spin at 10,000 g for 20 min. Wash the pellet with 500 μL of 70% ethanol, centrifuged for 10 min at 10,000 g. Once the supernatant is removed, the air-dried DNA pellet is ready to be given to a sequencing unit.

3.4. Validation of the TAIL-PCR Product

1. Examine the sequence that you obtained from the TAIL-PCR product. Identify the part of the sequence that corresponds to the end of the T-DNA element and the sequences corresponding to the cloning vector and trim.
2. Use the remaining sequence as a query in a Basic Local Alignment Search Tool (BLASTN) search against the *Arabidopsis* genome (http://www.Arabidopsis.org/Blast/index.html) to determine the chromosomal location of the T-DNA insertion.
3. Design two specific primers against the genomic sequence flanking the T-DNA insertion site, called "Primer LB" and "Primer RB." Carry out PCRs on DNA extracts with two different primer pairs: Primer LB and Primer RB to amplify the wild-type allele; GPTV-L3 and Primer RB to amplify the T-DNA flanking sequence. Add 1 µL of DNA from the selected line diluted at 5 ng/µL to a 10 µL PCR. Amplify also wild-type genomic DNA as a control. Each reaction contains 1 µL of each 10 µM primer, 0.2 unit of Taq-polymerase, 0.2 µL of 5mM dNTPs, 1 µL of 10× Buffer and 6.3 µL of sterile milliQ water. Perform the PCRs using the appropriate annealing temperature for your primers: 96 °C for 6 min, 35 cycles (95 °C for 1 min, annealing temperature for 1 min, 72 °C for 1 min) and 72 °C for 10 min. Separate 5 µL of the PCR products on a 2% agarose gel.

3.5. Sequence Analysis

Once you have determined that the TAIL-PCR fragment corresponds to the site of T-DNA insertion, the sequence can be used to determine the tagged gene.

1. Determine the orientation of the *GUS* gene within the chromosomal sequence.
2. Examine the annotated genome sequence to identify known or hypothetical genes that lie in the vicinity of the insertion site (*see* **Note 11**).
3. It may be possible to identify genes that have not yet been annotated by probing Northern blots and cDNA libraries with genomic DNA probes from near the site of insertion. The genomic DNA probes can be generated by PCR amplification. *GUS* can also be regulated by a gene that is located at a significant distance from the insertion site.

3.6. Does GUS Expression Accurately Reflect the Expression of the Tagged Gene?

After a candidate gene has been identified, its native expression pattern should be examined and compared to that of the *GUS* expression pattern in the trap line.

1. Examine the expression of the native gene in wild-type plants using Northern blot analysis, reverse transcription PCR (RT-PCR), or in situ hybridization. The choice of technique will depend on the distribution and apparent level of expression, as based on GUS activity. Northern blots and RT-PCR can give some indication of transcript distribution and abundance, but do not provide information about the

distribution of transcripts within a tissue or organ. In situ hybridization provides the best confirmation of the expression pattern of the endogenous gene, but may not be sensitive enough to allow detection of the very low levels of expression that can be reported by *GUS*, which is extremely sensitive *(19)*. Therefore, it may not always be possible to detect transcripts using in situ hybridization. It is also advisable to confirm the expression pattern of the endogenous gene, using promoter–reporter gene fusions.

2. Design primers to PCR amplify genomic sequences upstream of the translation start site. Regulatory sequences appear to be quite variable in size, so we recommend using genomic sequences that extend as far as the next upstream gene. In the majority of cases, 5′ sequences are sufficient for proper regulation, however in some cases, regulatory sequences reside 3′ to the transcribed region or in introns or exons within the gene (e.g., *see* **ref.** *20*).

3. Clone the promoter into a binary vector such as pCambia 1391Z, in front of a *GUS* reporter gene.

4. Introduce the binary vector into *Agrobacterium tumefaciens* strain GV3101 and transform the T-DNA construct into wild-type *Arabidopsis* plants using the floral dip method *(21)*. At least 10 transgenic lines containing independent promoter–reporter gene insertions should be generated and examined, as the genomic location of the transgene can result in variability in expression.

3.7. Is There a Mutant Phenotype Associated with the Insertion?

In addition to reporting the expression of adjacent chromosomal genes, a T-DNA insertion might also result in gene disruption. For this reason, the trap line should be examined for the presence of a mutant phenotype.

1. Grow plants that are homozygous for the T-DNA insertion and examine them for phenotypic abnormalities.

2. As the expression pattern can assist an educated guess in the search for a mutant phenotype, promoters that are specific to embryogenesis can eventually lead to phenotype affecting the embryo. Therefore, analyzing seeds in the silique of a heterozygote plants may uncover phenotypes affecting a quarter of the seeds corresponding to the homozygous progeny.

3. If a mutant phenotype is identified, it is important to verify that it is caused by the insertion of interest, as untagged mutations could be present in T-DNA populations. It is recommended to first determine whether the mutation is linked to the T-DNA element. Linkage can be examined in a population that segregates for both the mutant phenotype and the T-DNA. If possible, the analysis of a second allele available in T-DNA insertion collections provides very good supporting evidence. An alternative to demonstrate that disruption of a gene is responsible for the observed phenotype is to complement the mutation by introduction of a wild-type copy of the gene.

4. Ideally, complementation will be carried out by introduction of a full-length cDNA clone or a genomic clone that spans the entire gene. In either case, it is preferable

to use the gene's native promoter to drive expression, but a strong ubiquitously expressed promoter, such as the cauliflower mosaic virus *35S* promoter *(22)*, may also be suitable providing the mutation is expressed after the transition to heart stage. Use of a promoter such as *35S* is likely to result in expression at higher than normal levels and in ectopic expression. While this may allow complementation of the mutant phenotype, other phenotypic consequences may occur. If the mutation is recessive, and homozygous plants are fertile, then the complementation construct may be directly transformed into the mutant background. In this case, primary transformants can be directly examined for a rescue of the mutant phenotype. In the case of a mutation that causes lethality or decreased fertility, heterozygous plants or wild-type plants can be transformed. The complementation construct can be screened for co-segregation with the T-DNA trap or later be introduced into the mutant background by genetic crossing. It is preferable to analyze several insertion events to avoid bias due to positional effects.

4. Notes

1. We advise to make in advance a 4× stock solution without X-Gluc that is stable for months in the dark at room temperature. Ferricyanide and ferrocyanide are included in the solution to increase the specificity of staining. These chemicals promote the oxidative dimerization of the soluble monomer product of the reaction into a nondiffusible dimer, however, they also inhibit the GUS enzyme, resulting in decreased sensitivity. A concentration of 2 mM works well but can be adjusted as needed. The GUS staining solution can be used twice, unless a strong GUS activity had turned the stain blue, then it should be discarded. Chloramphenicol can be included to inhibit bacterial enzymatic activity resulting from contamination during the incubation.

2. For routine work and screening, we do not add Arabic gum to our Hoyer's medium.

3. Paraformaldehyde must be handled in a chemical fume hood while wearing appropriate gloves and goggles because it is highly toxic if it is inhaled or comes in contact with skin. To make the fixative, add 4 g of paraformaldehyde in 100 mL 1× PBS at pH 11 adjusted with 1M NaOH and preheated at 60 °C, stir to dissolve, cool on ice and adjust the pH to 7 with H_2SO_4 (not HCl), mix in 1 mL of triton X-100 and 1 mL of dimethyl sulfoxide to enhance penetration.

4. A different set of T-DNA primers should be used if working with a binary vector that differs from pBin19 in the T-DNA border sequences.

5. When the plants are grown in long days at 24 °C, embryogenesis lasts approximately 21 days, from fertilization to seed desiccation. The globular to heart stage transition occurs approximately 4–5 days. It is preferred to start tagging at the second or third flower as defects in fertilization are frequently observed in the first flowers. The seed coat of mature embryos represents a major barrier to X-Gluc entry, so the seed should be punctured with a needle under a dissecting scope.

6. This step is crucial to avoid inhibition of GUS activity by remaining acetone.

7. In some institutions, special permission is needed to order chloral hydrate because it is a controlled prescription substance. Chloral hydrate do not enter easily in the aqueous phase, several sparks of microwave can help it to dissolve. Before the embryo reaches the heart stage (approximately 5 days after fertilization), we advise to use the less disruptive Hoyer's light clarification (10 g Chloral hydrate, glycerol 3 mL, water 6 mL).

8. Glycerol, 50%, can be used as an alternative mounting medium. Glycerol mounting is not definitive; it offers the option of "fighting" air bubble formation without interfering with the quality of the observation. Glycerol-mounted slides can be sealed with nail polish and stored for long period of time.

9. Occasionally, no product is visible in the secondary reaction of a successful TAIL-PCR. If a clear product is present in the tertiary reaction, you may wish to determine its sequence. It is also common for more than one product to be amplified. In this case, you may see multiple bands in the secondary and tertiary TAIL-PCR products.

10. In our hands, this enzymatic degradation of primers gave excellent results over purification kits: no loss of DNA and good sequence read. Cloning the TAIL-PCR product prior to sequencing is required when multiple products are amplified in the TAIL-3 PCR. After gel purification, clone the PCR products using the pGEM-T vector system II (Promega) following the manufacturer's instructions. Isolate plasmids from the bacterial cultures using a QIAprep plasmid miniprep kit. Prior to sequencing, the authenticity and orientation of TAIL-PCR products can be determined by PCR amplification using the appropriate *T-DNA* primer in combination with either the T7 or the SP6 primer. This step will select for PCR products that did result from amplification of a flanking *T-DNA* sequence.

11. *GUS* expression may be observed from a T-DNA that is inserted so that *GUS* is in the opposite orientation to a gene in the region, and no genes in the correct orientation can be found. This may be due to the presence of cryptic promoter sequences *(23)*. Examples of expressed promoter- or gene-trap insertions in regions where no detectable transcribed gene is present have also been reported *(13,14,24)*. Reporter gene activation in these cases has been interpreted as being due to the activation of a cryptic promoter in the genome. However, it is also possible that a rare transcript is present in the vicinity and has been ignored in the annotation process. In particular, genes encoding small peptides and non-coding RNAs are likely to be underrepresented in the current annotation. We had successfully looked for non-orthodox transcripts by running the exon prediction software http://cbs.dtu.dk/services/NetPGene/.

Acknowledgments

The authors thank Dr. C. Carles for comments on the manuscript. Research related to promoter traps in embryos has been supported by CNRS funding and grants from the EU BRIDGE and EU EPEN programs.

References

1. Casadaban MJ, Cohen SN. Lactose genes fused to exogenous promoters in one step using a Mu-lac bacteriophage: in vivo probe for transcriptional control sequences. Proc Natl Acad Sci USA 1979;76:4530–4533.
2. Bellen HJ. Ten years of enhancer detection: lessons from the fly. Plant Cell 1999;1:2271–2281.
3. Springer PS. Gene traps: tools for plant development and genomics. Plant Cell 2000;12:1007–1020.
4. Stanford WL, Cohn JB, Cordes SP. Gene-trap mutagenesis: past, present and beyond. Nat Rev Genet 2001;2:756–768.
5. Bouche N, Bouchez D. Arabidopsis gene knockout: phenotypes wanted. Curr Opin Plant Biol 2001;4:111–117.
6. Laplaze L, Parizot B, Baker A, Ricaud L, Martiniere A, Auguy F, Franche C, Nussaume L, Bogusz D, Haseloff J. GAL4-GFP enhancer trap lines for genetic manipulation of lateral root development in Arabidopsis thaliana. J Exp Bot 2005;56:2433–2442.
7. Acosta-Garcia G, Autran D, Vielle-Calzada JP. Enhancer detection and gene trapping as tools for functional genomics in plants. Methods Mol Biol 2004;267:397–414.
8. Devic M, Hecht V, Berger C, Delseny M, Gallois P. An assessment of promoter trapping as a tool to study plant zygotic embryogenesis. C R Acad Sci Life Science 1995;121–128.
9. Topping JF, Wei W, Lindsey K. Functional tagging of regulatory elements in the plant genome. Development 1991;112:1009–1019.
10. Topping JF, Agyeman F, Henricot B, Lindsey K. Identification of molecular markers of embryogenesis in *Arabidopsis thaliana* by promoter trapping. Plant J 1994;5:895–903.
11. Dubreucq B, Berger N, Vincent E, Boisson M, Pelletier G, Caboche M, Lepiniec L. The *Arabidopsis* AtEPR1 extensin-like gene is specifically expressed in endosperm during seed germination. Plant J 2000;23:643–652.
12. Vielle-Calzada JP, Baskar R, Grossniklaus U. Delayed activation of the paternal genome during seed development. Nature 2000;404:91–94.
13. Fobert PR, Labbe H, Cosmopoulos J, Gottlob-McHugh S, Ouellet T, Hattori J, Sunohara G, Iyer VN, Miki BL. T-DNA tagging of a seed coat-specific cryptic promoter in tobacco. Plant J 1994;6:567–577.
14. Foster E, Hattori J, Labbe H, Ouellet T, Fobert PR, James LE, Iyer VN, Miki BL. A tobacco cryptic constitutive promoter, tCUP, revealed by T-DNA tagging. Plant Mol Biol 1999;41:45–55.
15. Casson SA, Chilley PM, Topping JF, Evans IM, Souter MA, Lindsey K. The POLARIS gene of Arabidopsis encodes a predicted peptide required for correct root growth and leaf vascular patterning. Plant Cell 2002;14:1705–1721.
16. Weigel D, Glazebrook J. Arabidopsis: A Laboratory Manual. CSH Laboratory Press, Cold Spring Harbor, NY. 2002.

17. Liu YG, Mitsukawa N, Whittier RF. Rapid sequencing of unpurified PCR products by thermal asymmetric PCR cycle sequencing using unlabeled sequencing primers. Nucleic Acids Res 1993;21:3333–3334.
18. Liu YG, Mitsukawa N, Oosumi T, Whittier RF. Efficient isolation and mapping of Arabidopsis thaliana T-DNA insert junctions by thermal asymmetric interlaced PCR. Plant J 1995;8:457–463.
19. Jefferson RA, Kavanagh TA, Bevan MW. GUS fusions: beta-glucuronidase as a sensitive and versatile gene fusion marker in higher plants. EMBO J 1987;6:3901–3907.
20. Sieburth LE, Meyerowitz EM. Molecular dissection of the AGAMOUS control region shows that cis elements for spatial regulation are located intragenically. Plant Cell 1997;9:355–365.
21. Clough SJ, Bent, AF. Floral dip: a simplified method for Agrobacterium-mediated transformation of Arabidopsis thaliana. Plant J 1998;16:735–743.
22. Benfey PN, Ren L, Chua NH. The CaMV 35S enhancer contains at least two domains which can confer different developmental and tissue-specific expression patterns. EMBO J 1989;8:2195–2202.
23. Cocherel S, Perez P, Degroote F, Genestier S, Picard G. A promoter identified in the 3′ end of the Ac transposon can be activated by cis-acting elements in transgenic Arabidopsis lines. Plant Mol Biol 1996;30:539–551.
24. Sivanandan C, Sujatha TP, Prasad AM, Resminath R, Thakare DR, Bhat SR Srinivasan. T-DNA tagging and characterization of a cryptic root-specific promoter in *Arabidopsis*. Biochim Biophys Acta 2005;1731:202–208.

11

Visualization of Auxin Gradients in Embryogenesis

Michael Sauer and Jiří Friml

Summary

Embryogenesis in *Arabidopsis thaliana* depends on the proper establishment and maintenance of local auxin accumulation. In the course of elucidating the connections between developmental progress and auxin distribution, several techniques have been developed to investigate spatial and temporal distribution of auxin response or accumulation in *Arabidopsis* embryos. This chapter reviews and describes two independent methods, the detection of the activity of auxin responsive transgenes and immunolocalization of auxin itself.

Key Words: Auxin in embryogenesis; *DR5*; fluorescent proteins; auxin transport; immunolocalization of auxin.

1. Introduction

The phytohormone auxin and its directional transport play an important role in plant embryogenesis. Studies involving mutants with auxin transport defects, auxin transport inhibitors, exogenous auxin application or manipulation of auxin homeostasis by ectopic expression of auxin biosynthesis genes have demonstrated a connection between local, polar auxin transport-dependent auxin distribution and proper embryo development *(1–4)*. However, by far, not every observed defect in embryogenesis is attributable to perturbed auxin distribution. Therefore, it can become necessary to investigate the actual auxin distribution in the embryo. To this end, several independent approaches to visualize auxin levels in embryos or other plant tissues have been developed, each with its own benefits and limitations. We will restrict this treatment to the two most commonly used approaches.

From: *Methods in Molecular Biology, vol. 427: Plant Embryogenesis*
Edited by: M. F. Suárez and P. V. Bozhkov © Humana Press, Totowa, NJ

1.1. Auxin Response Visualization with Auxin Responsive Transgenes

This is by far the most commonly used method for 'auxin visualization'. However, it is important to understand that it does not actually visualize the distribution of auxin. Instead, it reports the activity of a certain auxin response pathway, which depends on the concerted action of several factors. All necessary components of this pathway have to be present and active in the tissue of interest, for auxin to induce the expression of a reporter gene, which in turn is visualized.

This signalling pathway is now relatively well understood, for a comprehensive review, *see* **ref. (5)**. In brief, the mechanism involves the release of transcription factors from the auxin response factor (ARF) family from their repressors [Aux/indole-3-acetic acids (IAAs)] after auxin-induced degradation of the Aux/IAAs. Free ARFs then bind to auxin responsive elements (AREs) in promoters of auxin inducible genes, which then will be turned on or off. After identification of the conserved ARE motif in promoters of auxin-inducible genes *(6)*, artificial auxin responsive promoters (termed *DR5* or *DR5rev*) have been generated and fused to reporter genes, such a β-glucuronidase *(GUS)* which is detected by a chromogenic reaction *(7)* or green fluorescent protein *(GFP)*, which requires fluorescence microscopy for detection *(3)*.

The GFP-based variant allows for life imaging and also somewhat higher spatial resolution. However, it requires a fluorescence microscope equipped for GFP observation, and weak signals may not be detected easily. GUS-based constructs have the advantage of increased sensitivity, as the chromogenic reaction for its detection can be adjusted as desired. A disadvantage is potential diffusion of a reaction intermediate, which decreases spatial resolution, and in extreme cases leads to false results. Also, living material cannot be observed. Both GFP- and GUS-based transgenes share several problems. First, the dynamic range of these systems may be insufficient to accurately reflect actual auxin levels in your experiment. There needs to be expression of the reporter gene above a certain threshold for being able to detect its product at all. Auxin-induced expression below this threshold will escape detection. Likewise, you can loose dynamic fidelity in regions of high reporter activity, because the system may become saturated. Second, fast fluctuations in auxin concentrations cannot be accurately reflected, because the reporter genes need first to be expressed and their products properly folded in sufficient quantities, which delays the response. The half-life of the reporter gene products may be far greater than a transient auxin maximum, which would lead to an artifactually prolonged signal. Third, they are not specific for IAA, but respond to all biologically active auxins. This may or may not pose a problem for your experiment.

1.2. Immunocytochemical Visualization of Auxin

Immunolocalization of the most abundant natural auxin, IAA, is based on antibodies directed against carboxyl linked IAA. These antibodies have been shown to exhibit strong specificity for IAA *(8,9)* and have been used in a variety of plant species and tissues (e.g., *see* refs. *10–12*), as well as in *Arabidopsis* embryos *(3)*. However, it is not entirely clear if the antibodies exhibit the same specificity for IAA in situ as it has been shown in **refs.** *(8,9)*. This method is also much more time consuming and difficult than visualization of auxin responsive transgenes, but on the other hand, it does not require the existence of a complete auxin response machinery. For example, the *monopteros* mutant, which lacks ARF5 activity, fails to activate *DR5rev::GFP*, although undoubtedly there is auxin present in these embryos *(3)*. Also, the temporal accuracy is higher, as auxin itself is detected, not a transgenic protein with a specific half-life. However, we do not recommend this technique as a standard procedure due to the effort involved. Nevertheless, it can be useful as a second, independent approach to investigate auxin distribution in *Arabidopsis* embryos and other plant tissues.

2. Materials

2.1. Auxin Response Visualization with Fluorescent Auxin Responsive Reportergenes

1. Transgenic *Arabidopsis* plants harbouring a fluorescent protein (e.g., GFP) under control of an auxin responsive promoter (e.g., *GH3*, *DR5* or *DR5rev*).
2. Glycerol solution: 5% glycerol in water (*see* **Note 1**).
3. Fine syringe needles (0.3 mm diameter, e.g., insulin syringes), double-sided adhesive tape, microscope slides, cover slips, dissecting scope.
4. Epifluorescence microscope (widefield, confocal laser scanning or similar).

2.2. Auxin Response Visualization with β-Glucuronidase-Based Reporter Genes

1. Transgenic *Arabidopsis* plants harbouring GUS under control of an auxin responsive promoter (e.g., *GH3*, *DR5* or *DR5rev*).
2. Ice cold acetone, 80%, in water.
3. Washing buffer: 0.1 M Na_2HPO_4 at pH 7.0, 10 mM EDTA, 1% (v/v) Triton X-100, 1 mM $K_3[Fe(CN)_6]$ (potassium ferricyanide), 1 mM $K_4Fe(CN)_6$ (potassium ferrocyanide) in water.
4. Staining buffer: as washing buffer, but additionally with 1 mg/mL X-Gluc (5-bromo-4-chloro-3-indolyl-beta-D-glucuronic acid) (*see* **Note 2**).
5. Clearing solution: 66% (w/v) chloralhydrate, 24% (v/v) water, 10% (w/v) glycerol.

2.3. Direct Auxin Visualization by Immunolocalization

1. Fine syringe needles (0.3 mm diameter, e.g., insulin syringes), double-sided adhesive tape, coated microscope slides (e.g., Superfrost), cover slips 24 × 50 mm, dissecting scope, liquid repellent pen (e.g., Pap Pen, Sigma), humid chamber (a closed box lined with paper soaked in 1× phosphate buffered saline (PBS) buffer with a rack for horizontal placement of slides).

2. PBS, 10×, per 1 l: 80 g NaCl, 2 g KCl, 14.4 g Na₂HPO₄, 2.7 g KH₂PO₄, water to 1 l. Adujst pH to 7.4 with HCl. Prior to use, dilute to 1× PBS with water.

3. Prefixation solution: 3% (w/v) N-(3-Dimethylaminopropyl)-N′-ethylcarbodiimide hydrochloride (Sigma) in water, freshly prepared.

4. Fixation solution: 4% (w/v) paraformaldehyde (PFA) in 1× PBS (*see* **Note 3**) with 0.1% (v/v) Triton X-100. Solution can be stored for up to 3 months at –20°C in small aliquots.

5. Driselase solution: 2% (w/v) Driselase (Sigma) in 1× PBS. Driselase powder will not dissolve, mix the suspension vigorously, then let solids sediment while keeping the tube on ice. Use only the supernatant and prepare immediately before use.

6. Permeabilization solution: 3% (v/v) Igepal CA-630 (Sigma, equivalent to Nonidet P-40, which is no longer available), 10% (v/v) dimethyl sulfoxide in 1× PBS, freshly prepared.

7. Blocking solution: 2% (w/v) bovine serum albumin in 1× PBS, freshly prepared.

8. Primary antibody: mouse anti-IAA antibody (Agdia) (*see* **Note 4**) dilution 1:200 in 1× PBS.

9. Secondary antibody: alkaline-phosphatase conjugated anti-mouse antibody (e.g., goat anti-mouse IgG AP conjugate, adsorbed, Merck Calbiochem), dilution 1:1000 in blocking solution (*see* **Note 5**).

10. WesternBlue (Promega) alkaline phosphatase detection solution (*see* **Note 6**).

11. Detection buffer: 100 mM Tris–HCl pH 9.5, 50 mM MgCl₂, 100 mM NaCl in water.

12. Mounting solution: 90% (v/v) glycerol in 1× PBS.

3. Methods

3.1. Auxin Response Visualization with Fluorescent Auxin Responsive Reportergenes

1. Prepare a glass slide with double adhesive tape. Take siliques of the appropriate stage and place them on the double adhesive tape.

2. Under a dissecting scope, cut open the silique along the replum with syringe needles, stick the sides of the valves onto the tape and carefully scrape out the ovules with forceps.

3. Transfer the ovules to microscope slides into a drop of glycerol solution.

4. To release the embryos from their ovules, cover with a standard cover slip, then apply gentle to moderate pressure (e.g., with forceps or lead pencil) to squeeze out

the embryos (*see* **Note 7**). Alternatively, under a dissecting scope, preferentially with transmitted light, use syringe needles to cut open or tear the ovules, cover with a cover slip, then apply gentle pressure on the ovules (*see* **Note 8**).

5. Analyze the slides with a fluorescence microscope (epifluorescence widefield microscope, confocal laser scanning microscope, etc.). Use the appropriate excitation and emission filter settings for the fluorescent protein of your study (*see* **Note 9**).

3.2. Auxin Response Visualization with β-Glucuronidase-Based Reporter genes

1. Prepare a glass slide with double adhesive tape. Take siliques of the appropriate stage and place them on the double adhesive tape.
2. Under a dissecting scope, cut open the silique along the replum with syringe needles, stick the sides of the valves onto the tape and carefully scrape out the ovules with forceps.
3. Transfer the ovules to 1.5-mL microcentrifuge tubes with 80% acetone on ice, vacuum infiltrate for 5 min, then incubate at –20°C for 1 h.
4. Wash ovules two times for 10 min each with washing buffer at room temperature.
5. Transfer ovules to staining buffer, vacuum infiltrate for 5 min, then incubate at 37°C in darkness. The incubation time can range from several minutes to many hours and must be optimized experimentally (*see* **Note 10**).
6. Wash the ovules two times for 10 min each with water.
7. Transfer the ovules to a slide with a drop of clearing solution, cover with a cover slip and incubate for about 20 min. (*see* **Note 11**).
8. Analyze the slides preferentially with differential interference contrast optics (sometimes referred to as 'Nomarski' optics).
9. If the staining is too weak, too strong and/or diffuse, you can change the incubation time in step 5, X-Gluc and/or ferricyanide and ferrocyanide concentrations. (*see* **Note 12**).

3.3. Direct Auxin Visualization by Immunolocalization

1. Prepare a glass slide with double adhesive tape. Take siliques of the appropriate stage and place them on the double adhesive tape.
2. Under a dissecting scope, cut open the silique along the replum with syringe needles, stick the sides of the valves onto the tape and carefully scrape out the ovules with forceps.
3. Transfer the ovules to a microcentrifuge tube with 1 mL prefixation solution on ice, vacuum infiltrate for 5 min, then keep on ice for 1 h (*see* **Note 13**).
4. Remove the prefixation solution, wash once with 1× PBS (brief centrifugation at 8000 g can help to collect the ovules at the bottom of the tube), then add 500 μL of fixation solution. Vacuum infiltrate for 5 min, then keep at room temperature for 1 h.
5. Wash the ovules two times with 1× PBS.

6. Wash the ovules two times with water.

7. Place a small drop of water on the coated microscope slide and transfer the ovules on the microscope slide with a glass Pasteur pipette. Use ovules of about three siliques for one slide. Squash the ovules to release the embryos by gently pressing on them with the rounded back side of a lead pencil and let the sample dry out completely (2 h, overnight).

8. Mark the area around the dried material using the Pap Pen and rehydrate it with 250 μL 1× PBS for 5 min.

9. Remove the PBS and carefully pipette 200 μL of driselase solution onto the material. Incubate for 45 min at 37°C in a humid chamber.

10. Wash the slides two times in 1× PBS for 10 min each. (*see* **Note 14**).

11. Remove the PBS and carefully pipette 200 μL of permeabilization solution onto the material. Incubate for 1 h at room temperature in a humid chamber.

12. Wash the slides four times in 1× PBS for 10 min each (*see* **Note 14**).

13. Remove the PBS and carefully pipette 200 μL of blocking solution onto the material. Incubate for 1 h at room temperature in a humid chamber.

14. Remove the blocking solution and carefully pipette 100–200 μL of primary antibody solution onto the material. Incubate for 4 h at 37°C in a humid chamber (*see* **Note 15**).

15. Wash the slides four times in 1× PBS for 10 min each (*see* **Note 14**).

16. Remove the PBS and carefully pipette 100–200 μL of secondary antibody solution onto the material. Incubate for 3 h at 37°C in a humid chamber.

17. Wash the slides four times in 1× PBS for 10 min each (*see* **Note 14**).

18. Wash the slides in detection buffer for 10 min.

19. Remove detection buffer, carefully add 150 μL Western Blue solution and incubate at room temperature in a humid chamber in darkness. The duration of incubation must be determined experimentally, in our hands it ranged from several hours to overnight (*see* **Note 16**).

20. Once the desired staining strength is achieved, wash the slides with water for 10 min, then remove excess liquid and pipette 60–70 μL mounting solution onto the material. Gently cover with a 24- × 50-mm cover slip.

21. Observe the slides with brightfield or darkfield microscopy.

4. Notes

1. Throughout this chapter, water refers to bidistilled or reverse osmosis water.

2. Staining buffer can be stored for about 1 month at 4°C protected from light; however, X-Gluc must be added freshly to the buffer prior to use. Dissolve the required amount of X-Gluc in a small volume of dimethylformamide (DMF) first (approximately 5 μL DMF per mg of X-Gluc), then add to the staining buffer.

3. Increase pH of 1× PBS by dissolving a small amount of KOH pellets first, then add the weighed PFA powder. Afterwards, adjust pH back to 7.3 with HCl.

Caution: PFA is a harmful substance, handle powder under fume hood only and wear gloves.

4. There are several anti-IAA antibodies, the one Agdia sells refers to *(8)*. Another antibody with very similar characteristics has been described in **ref.** *(9)*, either of them should work.

5. For unknown reasons, the anti-IAA primary antibodies seem not to work well with fluorescently labelled secondary antibodies.

6. Alternatively, a different alkaline phosphatase detection system can be used, but we obtained the best results using this one.

7. This is best done with control under a transmitted light dissecting scope or, alternatively, under a normal dissecting scope, with a black surface underneath the slide. Practice with embryos of heart to late heart stage first, as you can see them popping out of the ovules quite easily. This will give you a feel about the amount of pressure to apply. Very early embryo stages are more difficult to see in a dissecting scope, in this case, just squash the ovules with moderate pressure. If results are unsatisfactorily, you can still apply more pressure afterwards, or use the alternative method, also given in **Subheading 3.1., step 4**.

8. The method of cutting the ovules open is more tedious and time consuming, but generally yields better results in terms of embryo morphology.

9. For GFP or eGFP, which are most often used, existing filter sets for FITC are compatible. Generally speaking, the excitation should be in the range of 450–490 nm (confocal users use 488 nm line of an argon laser), emission can be detected either with a long pass filter at 505 nm or a band pass from 500 to 540 nm. These values are approximate. With a long pass filter, be aware that you will detect chlorophyll as strong red fluorescence in older (heart stage and beyond) embryos.

10. The incubation time depends on the strength of GUS expression and can vary from several minutes to days. You can use the funiculus as an indicator for staining, as it should have a relatively strong signal.

11. Incubating in clearing solution at 37°C helps in the clearing process. If results are unsatisfactorily, prolong this step.

12. Higher concentrations of X-Gluc may reduce the incubation time and increase staining. Ferricyanide and ferrocyanide on the other hand inhibit GUS enzymatic activity. However, it is not recommended to reduce their concentration too much, as they catalyze the reaction from a diffusible, colourless intermediate to the indiffusible blue reaction product. If you observe diffusion, especially when incubation times are long (e.g., overnight), it can be necessary to increase the ferricyanide and ferrocyanide concentration to reduce diffusion.

13. Harvesting the ovules can take considerable amounts of time. Per slide, you will need ovules from about two to three siliques. We recommend collecting the ovules in the prefixation solution on ice, then do the vacuum infiltration for all samples at once.

14. Washing is best done in a box equipped with a slide holder, filled with the washing buffer under very gentle agitation. If you prefer, you can wash directly on the

slides by pipetting washing buffer on the material, but then you have to increase the number of washing steps by 50% or more, as the volume is much lower.

15. Alternatively, incubate overnight at 4°C, then an additional 1–2 h at 37°C.

16. You can check the progress of detection by washing away the substrate with detection buffer and subsequent microscopic evaluation. If the staining is not strong enough, add fresh Western Blue solution and incubate longer.

Acknowledgments

This work was funded by the Volkswagen Stiftung.

References

1. Liu C, Xu Z, Chua, N. H. Auxin polar transport is essential for the establishment of bilateral symmetry during early plant embryogenesis. Plant Cell 1993;5:621–630.

2. Shevell DE, Leu WM, Gillmor CS, Xia G, Feldmann KA, Chua NH. EMB30 is essential for normal cell division, cell expansion, and cell adhesion in Arabidopsis and encodes a protein that has similarity to Sec7. Cell 1994;77:1051–1062.

3. Friml J, Vieten A, Sauer M, Weijers D, Schwarz H, Hamann T, Offringa R, Jurgens G. Efflux-dependent auxin gradients establish the apical-basal axis of Arabidopsis. Nature 2003;426:147–153.

4. Weijers D, Sauer M, Meurette O, Friml J, Ljung K, Sandberg G, Hooykaas P, Offringa R. Maintenance of embryonic auxin distribution for apical-basal patterning by PIN-FORMED-dependent auxin transport in Arabidopsis. Plant Cell 2005;17:2517–2526.

5. Woodward AW, Bartel B. Auxin: regulation, action, and interaction. Ann Bot (Lond) 2005;95:707–735.

6. Ulmasov T, Liu ZB, Hagen G, Guilfoyle TJ. Composite structure of auxin response elements. Plant Cell 1995;7:1611–1623.

7. Ulmasov T, Murfett J, Hagen G, Guilfoyle TJ. Aux/IAA proteins repress expression of reporter genes containing natural and highly active synthetic auxin response elements. Plant Cell 1997;9:1963–1971.

8. Mertens R, Eberle J, Arnscheidt A, Ledebur A, Weiler E. Monoclonal antibodies to plant growth regulators. II. Indole-3-acetic acid. Planta 1985;166:389–393.

9. Leverone LA, Stroup TL, Caruso JL. Western Blot analysis of cereal grain prolamins using an antibody to carboxyl-linked indoleacetic acid. Plant Physiol 1991;96:1076–1078.

10. Moctezuma E. Changes in auxin patterns in developing gynophores of the peanut plant (Arachis hypogaea L.). Ann Bot (Lond) 1999;83:235–242.

11. Thomas C, Bronner R, Molinier J, Prinsen E, van Onckelen H, Hahne G. Immuno-cytochemical localization of indole-3-acetic acid during induction of somatic embryogenesis in cultured sunflower embryos. Planta 2002;215:577–583.

12. Hou ZX, Huang WD. Immunohistochemical localization of IAA and ABP1 in strawberry shoot apexes during floral induction. Planta 2005;222:678–687.

12

Intercellular Trafficking of Macromolecules During Embryogenesis

Insoon Kim and Patricia C. Zambryski

Summary

Plasmodesmata provide routes for communication and nutrient transfer between plant cells by interconnecting the cytoplasm of adjacent cells. A simple fluorescent tracer-loading assay was developed to monitor patterns of cell to cell transport via plasmodesmata specifically during embryogenesis. A developmental transition in plasmodesmatal size exclusion limit was found to occur at the torpedo stage of embryogenesis in *Arabidopsis*; at this time, plasmodesmata are downregulated, allowing transport of small (~0.5 kDa) but not large (~10 kDa) tracers. This assay system was used to screen for embryo defective mutants, designated *increased size exclusion limit of plasmodesmata* that maintain dilated plasmodesmata at the torpedo stage.

Key Words: *Arabidopsis*; embryogenesis; intercellular transport; plasmodesmata; size exclusion limit; ise; HPTS; dextran.

1. Introduction

Plant cells are connected via plasmodesmata, cytoplasmic channels that transverse plant cell walls. Plasmodesmata provide symplastic continuity between cells, facilitating communication and allowing co-ordination of growth and development. Plasmodesmata are plasma membrane-lined and have a core of modified endoplasmic reticulum (ER) in their center *(1,2)*. Transport through plasmodesmata is thought to occur primarily through the cytoplasmic space between the plasma membrane and modified ER *(3–5)*. An important measure of plasmodesmatal function is their size exclusion limit (SEL), the upper limit

From: *Methods in Molecular Biology, vol. 427: Plant Embryogenesis*
Edited by: M. F. Suárez and P. V. Bozhkov © Humana Press, Totowa, NJ

in the size of macromolecules that can freely diffuse from cell to cell (6–9). The aperture of plasmodesmata is regulated temporally, spatially and physiologically throughout the development of a plant, although SEL may be fixed in certain cell types or at specific stages of development (10–13).

Work to date on plasmodesmata has focused primarily on their ultrastructure (3), their roles during plant viral spread (14,15) and their SEL and functional state in different tissues or during different stages of post-germination development (16–21). Furthermore, recent work implies plasmodesmata have critical roles during complex morphogenesis, selectively allowing movement of some, but not all, transcription factors in the apical meristem (22–26). Plasmodesmata are also important players in plant defense, most dramatically illustrated in the recent wealth of data on gene silencing (27–32).

Despite the abundance of information on the structure and function of plasmodesmata, there are few reports that aim to identify the genes that control plasmodesmata architecture or regulatory components (33–35). Thus, we set out to specifically design a genetic screen to identify mutants with altered plasmodesmata. The screen begins with the premise that plants with defective plasmodesmata would likely have growth abnormalities and that homozygous mutants might not survive beyond early seedling stages. Therefore, we focused our genetic screen on *Arabidopsis* mutant lines segregating embryo-defective phenotypes. Such lines are maintained and propagated as heterozygotes, but their siliques bear homozygous mutant embryos.

2. Materials

2.1. Plant Growth and Harvest

1. Sunshine mix (Fisons Horticultural Inc.).
2. Vermiculite.
3. Plastic pots (e.g., 10 cm square) (Hummert).
4. Plastic trays and domes.
5. Hand sieves (mesh size = 0.425 mm).
6. Growth room (light intensity 120–150 μmol/m²s, humidity 25–75%, photoperiod of short days with 8 h light/16 h dark, and long days with 16 h light/8 h dark).

2.2. Fluorescent Probes

1. 8-Hydroxypyrene-1,3,6-trisulfonic acid (HPTS; MW 524 Da, Molecular Probes, H-348).
2. Fluorescein isothiocyanate (FITC)-conjugated 10-kDa dextrans (Sigma, FD-10S).
3. Sizing column with a 10 kDa cut-off (Amicon, YM-10 Microcon filters).
4. Murashige & Skoog (MS) media, 0.5× (pH 5.7) (Sigma).

2.3. Fluorescent Microscopy

1. Zeiss Axiophot epifluorescence microscope (Carl Zeiss, Inc.) equipped with mercury high-intensity light source and a color CCD camera (Optronics).
2. Scion Image 1.60 for CG7 software (Scion Corp.).
3. Zeiss FITC filter set (exciter BP470/20; dichroic beam splitter 510; emitter LP520).
4. Chroma FITC filter set (exciter BP480/30; dichroic beam splitter LP505; emitter BP535/40).
5. Zeiss 510 confocal laser scanning microscope system, equipped with Argon ion (488 nm) for HPTS and FITC fluorescence, and helium-neon lasers (543 nm), to monitor chlorophyll autofluorescence and the cytoplasmic localization of fluorescent tracers (Carl Zeiss, Inc.).
6. Adobe Photoshop software (Adobe).

2.4. Probe-Movement Assays on Embryos

1. #5 forceps.
2. Scissors.
3. Cover slip (22 × 22 mm) (Fisher).
4. Slide glass (Fisher).
5. Q tips.

3. Methods

3.1. Plant Growth

1. Sow seeds on the surface of a 1:1 mix of Sunshine mix and vermiculite (*see* **Note 1**).
2. Coverflats containing pots with a plastic dome and place at 4°C for 3 days for stratification (*see* **Note 2**).
3. To maintain high humidity during seed germination, maintain posts covered with a plastic dome for the first week after transfer to growth room from the cold room.
4. Grow plants at 22°C in growth rooms (*see* **Note 3**). Grow plants initially in short-day conditions, 8-h light/16-h dark cycle, for 1 weeks, to increase vegetative growth and subsequent inflorescence vigor.
5. Transfer plants to long-day conditions, 16-h light/8-h dark cycle, and bottom-water plants as needed.

3.2. Probe-Movement Assays on Embryos

1. Remove siliques from the stem by cutting petioles with scissors.
2. Open siliques by making a single slit with forceps.
3. Collect and place all seeds (typically 50–60 seeds per silique) on a microscope glass slide with 15–20 μL of fluorescent tracer solutions prepared.

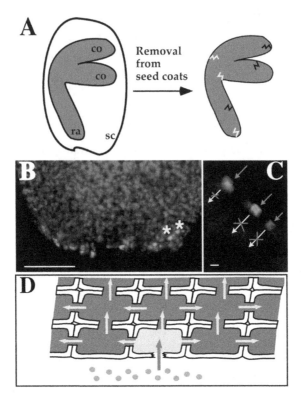

Fig. 1. Uptake of symplastic probes in cells of *Arabidopsis* midtorpedo embryos. (**A**) When embryos are released from their seed coats, physical damage occurs in a subset of cells. As a result, small regions of cell walls and plasma membranes are broken to a sub-lethal level to provide an initial entrance site for uptake of symplastic tracers such as 8-hydroxypyrene-1,3,6-trisulfonic acid (HPTS) and F-dextran, which do not cross plasma membranes. Jagged lines indicate the most common site of damage. co, cotyledon; ra, radicle; sc, seed coat. (**B**) A small number of cells at the base of the detached cotyledons from midtorpedo embryos are cytoplasmically loaded with 10-kDa F-dextran (asterisks); yet, further movement to neighboring cells does not occur as the plasmodesmata size exclusion limit (SEL) between cells is below 10 kDa. Scale bar, 50 μm. (**C**) A typical example of loaded cells in a region containing abrasion at the edge of the protodermal layer, marked as jagged lines in panel A. Individual cells in the protodermal layer take up 10-kDa F-dextrans (arrows) and show cytoplasmic localization of the probe. However, subsequent movement of the probe is inhibited (arrows with X). Scale bar, 5 μm. (**D**) A diagram shows how a partially broken cell wall and plasma membrane (jagged edge) may provide the initial entrance site for uptake of symplastic tracers, F-dextran or HPTS (circles). Further symplastic transport is then determined by the plasmodesmata SEL and the size of symplastic tracers introduced. Adapted from **ref.** *(38)*.

4. Place a cover slip over this solution and release embryos from their seed coats by slightly tapping the cover slip with a fingertip or a Q tip (*see* **Note 4**).

5. Incubate embryos in tracer solutions for 5 min at room temperature.

6. Wash out fluorescent tracers from under the cover slip with excessive 0.5× MS solution. Place a Q tip on one side of cover slip to absorb washing media and add 0.5× MS solution to the other side of cover slip. Repeat this procedure until no apparent tracer remains in the solution under the cover slip.

7. Immediately observe embryos by fluorescent microscopy. We initially examined probe movement in wild-type embryos to establish the foundation for the mutant screen (*see* **Fig. 1** and **3**)

3.3. Preparation of Fluorescent Probes

1. Prepare all fluorescent probes just before the movement assay.

2. Prepare HPTS (MW 524 Da) in 0.5× MS liquid culture media at 5 mg/mL. Usually prepare about 0.5 mL total as this will allow about 25 assays.

3. Prepare FITC-conjugated 10-kDa dextrans in 0.5× MS liquid culture media at 5 mg/mL. Purify 10-kDa F-dextrans using a sizing column with a 10 kDa cut-off to remove low molecular weight contaminants present in commercially prepared FTIC-conjugated dextrans.

4. Wrap the tubes containing fluorescent probes with foil to minimize bleaching by light.

3.4. Fluorescent Microscopy

1. Screening embryos for potential loading with fluorescent dyes is carried out by direct observation. Embryos loaded with fluorescent tracers are analyzed using a Zeiss axiophot epifluorescence microscope equipped with mercury high-intensity light source and a color CCD camera.

2. HPTS and FITC-dextran are analyzed using a Zeiss FITC filter set (exciter BP470/20; dichroic beam splitter 510; emitter LP520) revealing both green and red autofluorescent color, and a Chroma FITC filter set (exciter BP480/30; dichroic beam splitter LP505; emitter BP535/40) showing only green color from fluorescent probes.

3. Embryos that allow cell to cell movement of tracers appear green upon blue light excitation. Otherwise, embryos appear red due to chlorophyll auto-fluorescence .

4. Images are captured using Scion Image 1.60 for CG7 software.

5. For more detailed images, embryos are observed with a Zeiss 510 confocal laser scanning microscope system, equipped with Argon ion (488 nm) for HPTS and FITC fluorescence, and Helium Neon lasers (543 nm), to monitor chlorophyll auto-fluorescence.

6. Images are reconstructed using Adobe Photoshop software.

3.5. Isolation of Embryo-Defective Mutants with Increased SEL

1. To isolate *increased size exclusion limit of plasmodesmata* (*ise*) mutants, we purchased M2 ethyl methane sulfonate mutagenized seeds from Lehle Seeds (no. M2E-04-06, ecotype Landsberg *erecta*).
2. We visually screened 13,000 lines for altered embryo morphology at the torpedo stage of development and found 5000 lines segregating embryo-defective phenotypes.

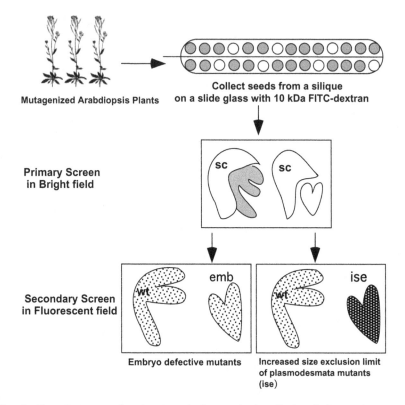

Fig. 2. Genetic screen for *increased size exclusion limit of plasmodesmata* (*ise*) mutants. Mature *Arabidopsis* siliques contain about 60 seeds; roughly half are drawn for simplicity. Gray seeds contain wild-type (wt) embryos, and white seeds contain defective embryos. Embryos are released from their seed coats (sc) and viewed under bright-field illumination. In the example diagrammed here, wt is gray and has reached the mid-torpedo stage, while the mutant is white and retarded in morphological development. Embryos are incubated with 10-kDa F-dextran for 5 min at room temperature, washed extensively, then viewed by fluorescent microscopy. wt and most morphologically defective embryo mutants (*emb*) emit red autoflorescence due to cholorophyll and are unable to transport 10-kDa dextran (left bottom panel). A small fraction of embryo defective lines (*ise*) take up and allow movement of 10-kDa F-dextran and thus exhibit green fluorescence (dark hatching, right bottom panel).

3. Among 5000 embryo defective lines, we identified 15 lines as putative mutants with altered SEL of plasmodesmata.
4. We initially characterized two mutants that showed a similar, dramatic increase in SEL (*ise1* and *ise2*). We also assayed a number of known mutants, including some from the collection of Meinke and colleagues *(36)*. One known mutant, *emb25* *(37)*, also showed an increased SEL, and was subsequently found allelic to *ise2*

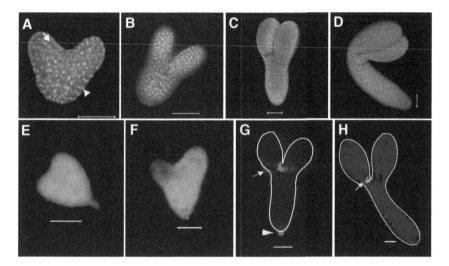

Fig. 3. Characterization of cell to cell transport in *Arabidopsis* wild-type embryos. Embryos at different stages of development are loaded with either 8-hydroxypyrene-1,3,6-trisulfonic acid (HPTS) (**A–D**) or 10-kDa FITC (**F**)-dextran (**E–H**). All cells in embryos allow the movement of HPTS, indicating that the embryo constitutes a single symplastic domain, from early heart (A), late heart (B), early torpedo (C), to mid torpedo (D) stages of embryo development. Cellular localization of HPTS shows tracer in the cytoplasm as well as the nuclei (arrowhead in A). In contrast, 10-kDa F-dextrans move only in early heart (E) and mid-heart (F) embryos. Eighty percent of early (G) and 100% of mid (H) torpedo embryos do not allow the movement of F-dextran. Instead, a small number of cells are loaded at the region where cotyledons join the hypocotyl (arrows in G and H) and the root tip (arrowhead in G); such loading occurs due to breakage of cell walls and plasma membranes during release of embryo from their seed coats. The images in panels A–D are optical sections captured by confocal laser scanning microscopy. Images in panels E–H were obtained by epifluorescent microscopy and are therefore less highly resolved. The images in panels A–F reveal the shapes of the embryos as green florescence (from HPTS or F-dextran) can be copied into gray scale. The images of the embryos for panels G and H cannot be seen except where a small amount of tracer has entered the embryos (arrows). As the original red auto-florescence from chlorophyll in the images in panels G and H does not copy into gray scale in a manner that can be distinguished from the green fluorescence of F-dextran, the outline of the embryos is shown as a white line. Scale bars, 50 μm. Adapted from **ref.** *(38)*.

by genetic complementation. The entire fluorescent probe loading procedure is summarized in **Fig. 2**. Stages of embryonic development of *Arabidopsis thaliana* shown in **Fig. 3** were adapted from **ref.** *(39)*.

4. Notes

1. Place three to five seeds per pot. Upon seed germination, leave three seedlings per pot so that each plant can have enough space to grow. When Metromix 300 or 250 soil is used (Scotts Co.), vermiculate is not necessary.
2. Fill the tray with water with a depth of 3–4 cm. This step helps to keep seeds moist during stratification to break the seed dormancy and to synchronize seed germination.
3. Commercial growth chambers, such as those manufactured by Conviron or Percival Scientific Inc., provide tight control of humidity, light intensity and duration. Light is provided by cool-white fluorescent bulbs or by incandescent lights. In contrast, in greenhouse, light is mainly provided by shaded sunlight and growth conditions are loosely controlled. Nevertheless, greenhouse-grown plants are healthy and are also excellent subjects for fluorescent probe uptake.
4. Probes are taken up initially into a subset of the cells most susceptible to damage during removal of the embryos from their seed coats. For example, due to the folding of the cotyledons, there are breaks at the cotyledon-shoot apical meristem junction. In addition, cells are broken by abrasion of the more exposed and protrusive root tip and at suspensor/root connection (*see* **Fig. 3G** and **H**). If the SEL is smaller than the size of the probe, further symplastic movement of probes does not occur.

Acknowledgments

We thank the NIH (GM45244) for support during the development of the described method and the Center for Biological Imaging for providing sound technical advice on fluorescence microscopy.

References

1. Tilney LG, Cooke TJ, Connelly PS, Tilney MS. The structure of plasmodesmata as revealed by plasmolysis, detergent extraction, and protease digestion. J Cell Biol 1991;112:739–747.
2. Robards AW. The ultrastructure of plasmodesmata. Protoplasma 1971;72:315–323.
3. Beebe DU, Evert RF. Photoassimilate pathway(s) and phloem loading in the leaf of *Moricandia arvensis* (L.) DC. (Brassicaceae). Int J Plant Sci 1992;153:61–77.
4. Overall RL, Wolfe J, Gunning BES. Intercellular communicationin Azolla roots. I. Ultrastructure of plasmodesmata. Protoplasma 1982;111:134–150.

5. Carr DJ. Plasmodesmata in growth and development. In: Gunning B, Robards RW, ed. Intercellular Communication in Plants: Studies on Plasmodesmata. New York: Springer-Verlag, Berlin, 1976:243–288.

6. Kempers R, Van Bel AJE. Symplastic connections beween sieve element and companion cell in the stem phloem of Vicia faba L. have a molecular exclusion limit of at least 10 kDa. Planta 1997;20:1195–1201.

7. Erwee MG, Goodwin PB. Characterization of the Egeria densa Planch. leaf symplast: inhibition of the intercellular movement of fluorescent probes by group II ions. Planta 1983;158:320–328.

8. Tucker EB. Translocation in the staminal hairs of Setcreasea purpurea. I. A study of cell ultrastructure and cell-to-cell passage of molecular probes. Protoplasma 1982;113:193–201.

9. Kempers R, Prior DAM, van Bel AJE, Oparka KJ. Plasmodesmata between sieve element and companion cell of extrafascicular stem phloem of *Cucurbita-maxima* permit passage of 3 kDA fluorescent probes. Plant J 1993;4:567–575.

10. Pickard BG, Beachy RN. Intercellular connections are developmentally controlled to help move molecules through the plant. Cell 1999;98:5–8.

11. Botha C, Cross R. Towards reconciliation of structure with function in plasmodesmata-who is the gatekeeper? Micron 2000;31:713–721.

12. Jackson D. Opening up the communication channels: recent insights into plasmodesmal function. Curr Opin Plant Biol 2000;3:394–399.

13. Citovsky V, Zambryski P. Systemic transport of RNA in plants. Trends Plant Sci 2000;5:53–54.

14. Carrington JC, Kasschau KD, Mahajan SK, Schaad MC. Cell-to-cell and long-distance transport of viruses in plants. Plant Cell 1996;8:1669–1681.

15. Lazarowitz SG. Probing plant cell structure and function with viral movement proteins. Curr Opin Plant Biol 1999;2:332–338.

16. Dolan L, Duckett CM, Grierson C, Linstead P, Schneider K, Lawson E, Dean C, Poethig S, Roberts K. Clonal relationships and cell patterning in the root epidermis of Arabidopsis. Development 1994;120:2465–2474.

17. Gisel A, Barella S, Hempel FD, Zambryski PC. Temporal and spatial regulation of symplastic trafficking during development in *Arabidopsis thaliana* apices. Development 1999;126:1879–1889.

18. Gisel A, Hempel FD, Barella S, Zambryski P. Leaf-to-shoot apex movement of symplastic tracer is restricted coincident with flowering in *Arabidopsis*. Proc Natl Acad Sci USA 1999;99:1713–1717.

19. Oparka KJ, Roberts AG, Boevink P, Santa Cruz S, Roberts I, Pradel KS, Imlau A, Kotlizky G, Sauer N, Epel B. Simple, but not branched, plasmodesmata allow the nonspecific trafficking of proteins in developing tobacco leaves. Cell 1999; 97:743–754.

20. Crawford KM, Zambryski PC. Subcellular localization determines the availability of non-targeted proteins to plasmodesmatal transport. Curr Biol 2000; 10:1032–1040.

21. Crawford KM, Zambryski PC. Non-targeted and targeted protein movement through plasmodesmata in leaves in different developmental and physiological states. Plant Physiol 2001;125:1802–1812.

22. Hantke SS, Carpenter R, Coen ES. Expression of floricaula in single layers of periclinal chimeras activates downstream homeotic genes in all layers of floral meristems. Development (Cambridge) 1995;121:27–35.

23. Kilby NJ, Fyvie MJ, Sessions RA, Davies GJ, Murray JA. Controlled induction of GUS marked clonal sectors in Arabidopsis. J Exp Bot 2000;51:853–863.

24. Kim JY, Yuan Z, Cilia M, Khalfan-Jagani Z, Jackson D. Intercellular trafficking of a KNOTTED1 green fluorescent protein fusion in the leaf and shoot meristem of *Arabidopsis*. Proc Natl Acad Sci USA 2002;99:4103–4108.

25. Kim J-Y, Yuan Z, Jackson D. Developmental regulation and significance of KNOX protein trafficking in *Arabidopsis*. Development 2003;130:4351–4362.

26. Wu X, Dinneny JR, Crawford KM, Rhee Y, Citovsky V, Zambryski PC, Weigel D. Modes of intercellular transcription factor movement in the Arabidopsis apex. Development 2003;130:3735–3745.

27. Baulcombe D. RNA silencing. Diced defence. Nature 2001;409:295–296.

28. Voinnet O, Lederer C, Baulcombe DC. A viral movement protein prevents spread of the gene silencing signal in Nicotiana benthamiana. Cell 2000;103:157–167.

29. Mourrain P, Beclin C, Elmayan T, Feuerbach F, Godon C, Morel JB, Jouette D, Lacombe AM, Nikic S, Picault N, Remoue K, Sanial M, Vo TA, Vaucheret H. Arabidopsis SGS2 and SGS3 genes are required for posttranscriptional gene silencing and natural virus resistance. Cell 2000;101:533–542.

30. Palauqui J-C, Balzergue S. Activation of systemic acquired silencing by localised introduction of DNA. Curr Biol 1999;9:59–66.

31. Voinnet O, Vain P, Angell S, Baulcombe DC. Systemic spread of sequence-specific transgene RNA degradation in plants is initiated by localized introduction of ectopic promoterless DNA. Cell 1998;95:177–187.

32. Palauqui J-C, Elmayan T, Pollien J-M, Vaucheret H. Systemic acquired silencing: transgene-specific post-transcriptional silencing is transmitted by grafting from silenced stocks to non-silenced scions. EMBO J 1997;16:4738–4745.

33. Evert RF, Russin WA, Bosabalidis AM. Anatomical and ultrastructural changes associated with sink-to-source transition in developing maize leaves. Int J Plant Sci 1996;157:247–261.

34. Ghoshroy S, Freedman K, Lartey R, Citovsky V. Inhibition of plant viral systemic infection by non-toxic concentrations of cadmium. Plant J 1998;13:591–602.

35. Chisholm ST, Mahajan SK, Whitham SA, Yamamoto ML, Carrington JC. Cloning of the Arabidopsis RTM1 gene, which controls restriction of long-distance movement of tobacco etch virus. Proc Natl Acad Sci USA 2000;97:489–494.

36. Franzmann LH, Yoon ES, Meinke DW. Saturating the genetic map of Arabidopsis thaliana with embryonic mutations. Plant J 1995;7:341–350.

37. Franzmann L, Patton DA, Meinke, DW. In vitro morphogenesis of arrested embryos from lethal mutants of *Arabidopsis thaliana*. Theor Appl Genet 1989; 77:609–616.

38. Kim I, Hempel FD, Sha K, Pfluger J, Zambryski PC. Identification of a developmental transition in plasmodesmatal function during embryogenesis in Arabidopsis thaliana. Development 2002;129:1261–1272.

39. Meinke DW. Seed developement in Arabidopsis thaliana. In: Meyerowitz EM, ed. Arabidopsis. Cold Spring Harbor, NY: Cold Spring Harbor Laboratory Press, 1994:253–295.

13

Immunolocalization of Proteins in Somatic Embryos

Applications for Studies on the Cytoskeleton

Andrei P. Smertenko and Patrick J. Hussey

Summary

Plant embryogenesis requires a tight balance between cell proliferation and differentiation. In animals, embryogenesis is dependent on cell migrations, which is in contrast to plant embryogenesis where the rigid cell wall precludes migration. Therefore, plants have to position cells correctly by defining the direction of the division plane during proliferation and control cell shape by local cell expansion. Both these processes are reliant on the organization and dynamics of the cytoskeleton – actin filaments and microtubules. In previous work *(7)*, we have shown that differentiation of the embryo suspensor is accompanied by reorientation of microtubules from random to transverse and reorganization of actin filaments from a fine filamentous network to bundled longitudinal cables. Here, we describe the technique for visualization of cytoskeletal components including actin filaments, microtubules and their associated proteins during the development of plant embryos in whole-mount specimens.

Key Words: Embryo development; microtubules; actin filaments; antibodies; immunolocalization.

1. Introduction

Plants have two known cytoskeletal elements: microtubules and actin filaments. Microtubules are dynamic structures formed from the polymerization of tubulin, a heterodimeric complex consisting of an α- and a β-tubulin subunit *(1)*. Another component of microtubules, microtubule-associated proteins, control the

From: *Methods in Molecular Biology, vol. 427: Plant Embryogenesis*
Edited by: M. F. Suárez and P. V. Bozhkov © Humana Press, Totowa, NJ

polymerization and the organization of microtubules *(2)*. Microtubule organization changes dramatically through the cell cycle forming four distinct arrays: two cortical microtubule arrays, the interphase cortical array and the preprophase band which appears at the G2/M phase transition, the mitotic spindle, and the phragmoplast which appears late in anaphase and in telophase *(3)*. The interphase cortical array, the preprophase band and phragmoplast are unique to plants. Cortical microtubules are located in the thin layer of cytoplasm underlying the plasma membrane and provide tracks for the movement of cellular components including the cellulose synthase complex and therefore coordinates cell wall biogenesis and consequently the direction of the cell expansion *(4)*. It has been noted that the direction of plant cell expansion is perpendicular to the orientation of cortical microtubules. The pre-prophase band defines the position and plane in which the cell will divide and in this way determines the architecture of plant tissues, the mitotic spindle separates daughter chromosomes and the phragmoplast is responsible for the formation of the cell plate at the position defined by the preprophase band between the daughter cells. Actin filaments (F-actin) are highly dynamic structures formed from monomeric actin or G-actin *(5)*. Actin filaments are present both in interphase and mitotic cells and can form a fine cortical network and thick, bundled transvacuolar cytoplasmic cables. Actin filaments are essential for cell expanison, cytoplasmic streaming and vesicular movement *(5)*.

Visualization of proteins in vivo yields important information about their role in cellular events. Immunohistochemistry was the first method to study localization of specific proteins in fixed cells and tissues; however, the utilization of fluorescent proteins as chimeric fusions has revolutionized the field of bioimaging as this technology allows the visualization of proteins in living cells and within a context of cellular events. Both techniques have their advantages and disadvantages. Immunohistochemistry requires fixation and permeabilization of the specimens, a process that kills the cells and that may cause some re-distribution of the proteins. The specimen is subsequently incubated with a specific antibody against the protein of interest (known as the primary antibody). This is followed by the secondary antibody conjugated to a fluorochrome, which binds to the primary antibody (because two antibodies are used this technique is called indirect immunofluorescence microscopy). This fluorochrome emits light when excited by the correct wavelength revealing the localization of the protein. It is now relatively easy to generate recombinant proteins, and there are numerous companies that sell expression systems (prokaryotic, eukaryotic and cell-free systems) and also generate antibodies from recombinant proteins. This technique has the added advantage of identifying the location of post-translationally modified proteins provided the antibodies only detect the modified version. In using fluorescent proteins as tags, the protein of interest is expressed in frame with the tag under the control of a constitutive, inducible or tissue-specific promoter. The fluorescent tag will emit light when excited with its specific wavelength. Although, first considered to be a non-invasive technique, the use of fluorescent protein tags does have some disadvantages. Firstly, fusion proteins are often over-expressed and driven by non-native promoters with the result that excess active protein causes further aberrant localization and/or affects the viability of the host cell *(6)*. Secondly, the fusion protein may not have the same activity as the native protein. For these reasons, it is prudent to use both fixed and live cell imaging technologies when studying the localization of proteins in vivo.

The major consideration in immunolocalization studies is the source of primary antibody. The best antibodies are going to be those that are made specifically to your protein of interest. However, in the interest of speed and economy, it is possible to use antibodies to proteins from other organisms provided the protein has a good conservation of structure. For example, the tubulin primary structure is evolutionary conserved making it possible to use antibodies made against yeast (YOL34) or animal (DM1A) tubulins for the visualization of microtubules. The primary structure of the plant microtubule associated proteins however is not always sufficiently conserved and may require preparation of the specific antibodies. Visualization of plant actin filaments is limited by poor cross-reactivity of antibodies made against animal actin with plant actin and lack of commercial antibodies against plant actin. A commercially available monoclonal antibody (clone C4) to animal actin has been successfully used for the staining of actin filaments in plants and is available from ICN (cat. no. 691002). Actin filaments in plants can also be stained using derivates of the fungal alkaloid phalloidin covalently linked to fluorochromes.

Before starting an immunolocalization experiment, it is important to test the specificity of the antibody and to determine the correct dilution of the antibody to be used. This can be done by using the antibody to probe one-dimensional or two-dimensional gel western blots of total protein extracts isolated from the organ/tissue where immunolocalization is intended. Where an antibody produces no signal on the western blot or generates a strong background, it should not be used and an alternative antibody has to be chosen.

Here, we describe a technique for the visualization of cytoskeletal proteins in the embryos of Norway spruce *Picea abies (7)*. The first steps involve fixing the material and permeabilizing the cells so that the antibodies can attach to their target molecules. The cells are then settled onto cover slips and incubated with the primary and secondary antibodies or with rhodamine-conjugated phalloidin if actin is to be stained in this way. At the end, the samples are mounted in a medium, which both stabilizes the fluorescent signal during imaging and preserves the specimen.

2. Materials

All chemicals used are of ANALAR grade or Molecular Biology grade. Deionized water is used throughout, and all incubations are at room temperature unless otherwise stated. Good laboratory practice is adhered throughout.

2.1. Tissue Culture of Embryogenic Cell Lines

1. LP medium *(8,9)*: 18.8 mM KNO_3, 15 mM NH_4NO_3, 1.5 mM $MgSO_4$, 2.5 mM KH_2PO_4, 3.0 mM $CaCl_2$, 50 μM Fe-EDTA, 10 μM Zn-EDTA, 10 μM $MnSO_4$, 10 μM H_3BO_3, 0.1 μM Na_2MoO_4, 0.01 μM $CuSO_4$, 0.01 μM $CoCl_2$, 4.5 μM KI, 4.9 μM pyridoxine-HCl, 16.3 μM nicotinic acid, 26.7 μM glycine, 15 μM thiamine-HCl, 0.5 mM *myo*-inositol, 1 mM D-glucose, 1 mM D-xylose, 1 mM L-arabinose, 0.56 μM L-alanine, 0.13 μM L-cysteine-HCl, 0.06 μM L-arginine, 0.08 μM L-leucine, 0.06 μM L-phenylalanine, 0.06 μM L-tyrosine. The medium is diluted to half-strength.
2. Growth room maintaining 20 °C.
3. Orbital shaker.
4. Laminar flow hood.

2.2. Protein Extraction

1. Mortar (about 10 cm in diameter) and pestle.
2. Liquid nitrogen.
3. Heat block set at 90–100 °C or boiling water bath.
4. Bench top centrifuge.

2.3. Sodium Dodecyl Sulphate Poly-Acrylamide Gel Electrophoresis

1. Sodium dodecyl sulphate–poly-acrylamide gel electrophoresis (SDS–PAGE) minigel kit. We use the kit from Atto Bioscience and Biotechnology Corporation (http://www.atto.co.jp).
2. Two times concentrated SDS–PAGE sample buffer *(10)*: 2 mL of water, 5 mL of 0.25 M Tris–HCl, pH 6.8, 2 mL of glycerol, 1 mL of β-mercaptoethanol and 0.4 g of SDS powder. Add 1 mg of bromophenol blue. Mix well and keep at –20 °C.
3. Acrylamide/Bis-acrylamide stock solution, 30%: 29.2 g acrylamide, 0.8 g of bis-acrylamide, water up to 100 mL. Mix and filter the solution through 40 μM nitrtocellulose membrane, store at +4 °C (*see* **Note 1**).
4. Stacking gel mixture. 5 mL of stacking gel mixture contains 2.5 mL of 0.25 M Tris–HCl, pH 6.8, 0.67 mL of acrylamide stock (final concentration 4% w/v), 1.75 mL of water, 50 μL of 10% (w/v) SDS solution, 50 μL of 10% ammonium persulphate solution (*see* **Note 2**).
5. Resolving gel mixture. 10 mL of resolving gel mixture contains 5 mL of 0.75 M Tris–HCl, pH 8.8, 100 μL of 10% (w/v) SDS solution, 100 μL of 10% (w/v) ammonium persulphate solution, appropriate amount of acrylamide stock and water upto 10 mL.

6. Gel running buffer: 0.025 M Tris, 0.19 M glycine and 0.1% (w/v) SDS.
7. Gel staining solution: 35% (v/v) ethanol, 7% (v/v) acetic acid, 1.25% (w/v) Coomassie Brilliant Blue R-250. To make the gel staining solution, mix ethanol with acetic acid, then add Coomassie Brilliant Blue R-250 and mix for several hours (this stage can be done overnight). Filter the solution through a paper filter and add water to the required volume. This solution can be re-used at least five times providing that gel staining is done in a closed container.
8. Gel destaining solution: 35% (v/v) ethanol, 7% (v/v) acetic acid.

2.4. Western Blotting

1. Gel transfer buffer: 0.025 M Tris, 0.19 M glycine, 0.002% (w/v) SDS, 20% (v/v) methanol.
2. Protein blotting quality Nitrocellulose membrane. We use Electran Nitrocellulose membrane from VWR, cat. no. 436106D.
3. Filter paper, cut to the size of the gel.
4. Tris Buffered Saline supplemented with Tween 20 (TBST) buffer: 20 mM Tris–HCl, pH 7.4, 150 mM NaCl, 0.02% (v/v) Tween 20.
5. Amidoblack solution: 1% (w/v) of amido black 10B (Naphtol blue black) in 0.6% (v/v) acetic acid. The solution is stable for 1 year. Dilute 100 times before use.
6. Mini-incubation trays for membrane strips, BioRad, cat. no. 1703902.
7. Blocking solution: TBST supplemented with 5% (w/v) fat-free powdered milk.
8. Secondary antibody conjugated with horseradish peroxidase (or alkaline phosphatase) is stored and diluted according to manufacturers' recommendations.
9. Enhanced chemiluminescence kit for western blot development.

2.5. Preparation of Poly-L-Lysine-Coated Cover Slips

1. Prepare chromosulfuric acid by adding 10 g of $K_2Cr_2O_7$ to 100 mL of concentrated H_2SO_4 in a 500-mL Duran bottle, swirl gently several times and leave overnight. Swirl several times again in the morning. Dissolving $K_2Cr_2O_7$ in H_2SO_4 (leads to the chemical reaction $K_2Cr_2O_7 + H_2SO_4 = 2CrO_3 + K_2SO_4 + H_2O$) results in the production of chromium trioxide, a very powerful oxidizing, airborne and extremely toxic agent. Prepare and handle chromosulfuric acid in the fume cupboard. The chromosulfuric acid has a dark brown colour and can be stored at room temperature in a fume cupboard in a closed glass container for several years. It can be reused until the colour changes to dark green.
2. 18 × 18 or 22 × 22 mm No. 1 cover slips (e.g., cat. no. 631 0124 or 6310120, VWR).
3. Poly-L-lysine (MW 70,000) solution, 1 mg/mL in water. Store at –20 °C in 1 mL aliquots. Each aliquot survives several freeze/thaw cycles.

2.6. Fixation and Permeabilization of the Embryos

1. 10× PEM (PIPES, EGTA, MA) buffer: 500 mM Piperazine-N,N'-bis(2-ethansulphonic acid) (PIPES)-NaOH, pH 6.8, 50 mM ethylene-glycol-bis (β-aminoethylether)-NNN´N´-tetra-acetic acid (EGTA), 20 mM MgSO₄. Store at –20 °C. This buffer can withstand multiple freeze/thaw cycles (*see* **Note 3**).
2. Phosphate-buffered saline (PBS) buffer: 2 mM KH₂PO₄, 10 mM Na₂HPO₄, 150 mM NaCl, pH 7.2.
3. Protease inhibitor stocks. 100 mM phenylmethylsulphonylfluoride (PMSF; 100 times stock solution in methanol), 40 mM leupeptine (1000 times stock solution in water), 20 mM pepstatin A (1000 times stock in methanol). Store at –20 °C.
4. Cell wall digesting enzyme mixture: 15 mM MES-NaOH, pH 5.0, 0.4 M mannitol, 5 mM EGTA, 1% (w/v) macerozyme, 0.8% (w/v) pectolyase. Split the solution into 5 mL aliquots and store at –20 °C. Do not reuse the aliquots. Before use, add PMSF, leupeptine and pepstatin A to a final concentration of 1 mM, 40 µM and 20 µM respectively.

2.7. Probing with the Antibodies and Actin Staining

1. PBS/BSA/SA buffer: PBS supplemented with 2% (w/v) bovine serum albumin (BSA) and 0.02% (w/v) sodium azide (NaN₃). Use high purity, protease-free BSA for this buffer (e.g., molecular biology grade) as contamination with proteases may cause degradation of the antibodies.
2. Anti-tubulin antibody, DM1A (cat. no. T9026, Sigma). Working dilution for probing western blots is 1:500 and for immunofluorescence 1:200.
3. Anti-mouse fluorescein isothiocyanate conjugates (cat. no. 715-095-150, Jackson Immunoresearch). Working dilution 1:200.
4. DAPI/PBS/SA buffer (DNA stain): PBS supplemented with 0.02% (w/v) NaN₃ and 40 ng/mL 4′,6-diamidino-2-phenylindole (DAPI).
5. Rhodamine phalloidin conjugate from Molecular probes (cat. no. R-415).
6. Slide mounting medium (e.g., Vectashield from Vector Laboratories).

3. Method

3.1. Tissue Culture of Embryogenic Cell Lines

1. Prepare half-strength LP medium, add D-Sucrose to a final concentration of 30 mM, adjust pH to 5.8 ± 0.1 and aliquot 47 mL of the medium into 250-mL Erlenmeyer flasks.
2. Autoclave the medium at 121 °C for 15 min.
3. When the medium has cooled to room temperature, add filter-sterilized L-glutamine to a final concentration of 3 mM, 2,4-dichlorophenoxyacetic acid to a final concentration of 9.0 µM and N6-benzyladenine to a final concentration of 4.4 µM (steps 3 and 4 must be performed aseptically in a flow hood).
4. Inoculate 3 mL of the older culture into 47 mL of fresh medium every 1–2 weeks.
5. Grow cells at 20 °C on an orbital shaker set at 100 rotations per minute.

3.2. Protein Extraction

1. Filter 10–40 mL of the liquid embryogenic culture though 50-μM nylon mesh to collect 0.5–1 mL of packed cells. Blot the nylon filter onto the paper towel to remove excess media.
2. Preheat SDS–PAGE sample buffer to 90–100 °C in a dry block or water bath in the fume cupboard.
3. Immerse the collected cells in a mortar filled with liquid nitrogen using a spatula.
4. Pre-chill the pestle in liquid nitrogen and grind the embryos to a fine powder.
5. In the fume cupboard, add 1 volume of hot SDS–PAGE sample buffer to 4–5 volumes of cell powder, vortex until the buffer mixes homogeneously with the cell powder. Incubate at 90–100 °C for 5 min.
6. Centrifuge for 5 min at 16,000 g and transfer the supernatant to a fresh plastic tube. Keep at –20 °C until needed .

3.3. Sodium Dodecyl Sulphate–Poly-Acrylamide Gel Electrophoresis

1. Wash gel glass plates with detergent, rinse with deionized water and ethanol. Air dry.
2. Assemble the glass plates as recommended by the manufacturer.
3. Prepare the resolving gel mix (*see* **Note 4**), add 5 μL of N,N,N′,N′-tetramethylethylenediamine (TEMED) per 10 mL of mixture and fill to three quarters the glass plate height. Overlay the top of the solution with ethanol.
4. Wait for the gel to polymerize (an edge of the polymerized gel becomes visible below the interface between the gel mixture and ethanol. A slight tilting of the glass plate assembly helps to determine whether the gel is set).
5. Wash the surface of the gel with distilled water. Invert the plates to drain off the water.
6. Prepare the stacking gel mixture. Rinse gel combs in distilled water, dry and place them into the glass plate assembly. Add 10 μL of TEMED to 5 mL of the stacking gel mixture and quickly load the solution into a syringe with a 22-gauge needle. Fill the gel plate assembly up to the edge of the glass plates. Remove air bubbles trapped underneath the comb by gently shaking the glass plate assembly. Top up the level of the stacking gel mixture after removing the air bubbles if necessary. The stacking gel sets within several minutes, so work quickly.
7. When the stacking gel sets, gently remove the comb and rinse the wells with distilled water using a syringe.
8. Set the electrophoresis as advised by the manufacturer.
9. First assess the quality of the protein extraction. Load protein molecular weight markers followed by 5, 10, 20 μL of the protein extract(s) into adjacent wells.
10. Run the gel as recommended by the manufacturer until the bromophenol blue line at the front of the run reaches the bottom edge of the gel.
11. Incubate the gel in the staining solution for 30 min and then wash three times, 20 min each, in the destaining solution.

12. Analyse the image. To assess the amount of protein required for western blotting in the following section (*see* **Subheading 3.4**), compare the intensity of the marker controls with the different loadings of the extract. As an approximate guide, choose the loading that shows clear bands stained with Coomassie blue.

3.4. Western Blotting

1. Prepare the gels as outlined in the **Subheading 3.3.** Load samples in all lanes leaving one for the markers. Run the gel as described above.
2. After electrophoresis, take the gel out of the glass plates and wash in the gel transfer buffer for 5 min.
3. Equilibrate one transfer membrane and four filter papers in the transfer buffer.
4. Arrange the components of the gel transfer system in the following order starting from the side which will face the plus electrode: two pieces of filter paper, the membrane, gel, two pieces of filter paper. Remove the bubbles by rolling a pipette over the sandwich. Set the transfer as recommended by the manufacturer (if protein transfer is done in the tank and no instructions are available use 50 V for the 2-h transfer and 20 V for the overnight transfer).
5. Remove the membrane, briefly wash in distilled water and stain in the amido black solution until the protein bands become visible. Afterwards rinse the membrane again in water and air dry on a peace of filter paper (*see* **Note 5**).
6. When dry, attach a strip of sticky tape along the top edge of the membrane covering 2–3 mm from the top of the membrane and fix the remaining part of the sticky tape on a piece of clean card paper. This will keep the membrane in place during subsequent cutting.
7. Cut the membrane through the middle of each lane and between the lanes to make strips. Label each strip at the bottom end with a sharp pencil. Cut three strips for each antibody to be tested. Cut along the edge of the sticky tape to release the strips.
8. Place strips in separate wells of a mini-incubation tray, rinse with water, add blocking solution and incubate the strips for 20 min. Gently shake or rock the strips throughout all subsequent incubations.
9. Prepare three dilutions of each antibody to be tested in the blocking solution, 1:100, 1:500 and 1:1000 (*see* **Note 6**).
10. Add 0.5 mL of diluted primary antibody to each strip and incubate for 1 h (*see* **Note 7**).
11. Wash the strips three times for 10 min each in TBST.
12. Incubate the strips with the secondary antibody diluted in blocking buffer as recommended by the manufacturer for 40 min.
13. Wash the strips in TBST three times for 10 min each.
14. Prepare the substrate for detecting the secondary antibody as recommended by the manufacturer allowing 1 mL of the substrate per 50 cm^2 of membrane.

15. Perform the following steps in the dark room if X-ray film is used for the detection (*see* **Note 8**); make sure that everything is set up for the development of the film before you start.

16. Place a piece of filter paper in a Petri dish or a tray, moisten the filter paper with water; remove the air bubbles trapped underneath the paper, put a peace of parafilm on the filter paper avoiding any bubbles underneath the parafilm. Arrange all strips next to each other protein side up on the parafilm.

17. Remove excess buffer from the strips by tilting the Petri dish so that the buffer accumulates at one end of the strips and gently blot the buffer with a peace of filter paper or paper towel.

18. Apply the substrate solution directly onto the strips; spread the substrate evenly. Allow the reaction to proceed for the time specified by the kit manufacturer.

19. Remove excess substrate solution as described in **step 16**. Pick up the parafilm with the strips and place it between two peaces of transparent plastic film inside an X-ray film cassette.

20. Take 1-min and 5-min exposures; develop the film. Increase the exposure time to 10–20 min or decrease the exposure to 10–30 s if the signal is too weak or too strong respectively (*see* **Note 9**).

21. The optimal dilution of the antibody gives the strongest signal with minimum background (*see* **Fig. 1**; *see* **Note 10**). Use this dilution and two times lower dilution for immunofluorescence experiments.

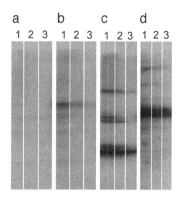

Fig. 1. Western blots of the same plant total protein extract probed with four different antibodies. These demonstrate the type of results that can be obtained. The antibodies were diluted 1:250 (lane 1), 1:500 (lane 2) and 1:1000 (lane 3). (**A**) no signal, some background is visible at lowest dilution; (**B**) the strong signal at lowest dilution is accompanied by strong background; however, at higher dilutions, both response and background weaken; (**C**) strong signal, but a lot of non-specific bands are also detected; (**D**) the ideal response is when a strong signal persists at high dilutions whilst the background disappears. If an antibody produces signal similar to one shown in panel (**D**), dilute antibody 1:500 and 1:1000 for immunolocalization studies.

3.5. Preparation of Cover Slips

3.5.1. Washing Cover Slips

1. Place 100–200 cover slips into a 2.5-l Erlenmeyer flask and add general-purpose laboratory detergent. Gently rotate the flask using an orbital or reciprocal shaker for 30 min. Set the speed of the shaker so that the cover slips are moving slightly. Pour off the detergent.
2. Add 0.5 l of deionized water, swirl gently and pour the water off keeping the cover slips in the flask. Repeat five times. Pour off the final wash and invert the flask on a piece of paper towel. Let the water drain down.
3. Add chromosulfuric acid, swirl gently and leave O/N in the fume cupboard. Check that all cover slips are submerged in the chromosulfuric acid.
4. Wash the cover slips in deionized water several times and then leave on the shaker, changing water every 30 min until no yellow colouring remains in the flask.
5. Air-dry the cover slips on a paper towel. For best results, separate attached cover slips and arrange them individually on the clean paper towel (*see* **Note 11**).
6. Put the dried cover slips in ethanol : ether (1:1) solution in an air-tight jar. The cover slips can be used after 12 h and can be stored in this solution for any time providing that ethanol : ether solution stays clean and is topped up regularly.

3.5.2. Coating Cover Slips with Poly-L-Lysine

1. Take cover slips out of the storage solution, place on a fresh piece of filter paper and air-dry (*see* **Note 11**).
2. Apply 70 μL or 100 μL of poly-L-lysine solution per 18 × 18-mm or 22 × 22-mm cover slip respectively. Spread the solution evenly over the cover slips. If the washed cover slips were stored and handled correctly, the poly-L-lysine solution will not retract to the centre of the cover slips. If poly-L-lysine solution is retracted to the centre of the cover slip, the adhesion of cells to this cover slip is not likely to be good.
3. Keep poly-L-lysine solution on the cover slips for 20 min.
4. Pick cover slips up by the corner with tweezers and wash in distilled water, place on a filter paper, poly-L-lysine treated side up, and air-dry (*see* **Note 12**).

3.6. Fixation and Permeabilization of the Embryos

1. Defrost 10× PEM buffer and set up a water bath at 60 °C.
2. Weigh 740 mg of paraformaldehyde and place into a 50-mL centrifuge tube (*see* **Note 13**).
3. Add 10 mL of water, close the tube and vortex until the paraformaldehyde makes a homogeneous suspension (*see* **Note 14**).
4. Place tube into a 60 °C water bath for 5 min stirring several times during incubation and then add 50 μL of 0.1 M NaOH, stir the tube and place back into the water bath and stir occasionally until the solution clears (*see* **Note 15**).

5. Immediately add 2 mL of 10× PEM buffer and keep the tube at room temperature.
6. Add 100 μL of Triton X-100 and stir or pipette gently to dissolve.
7. Cool the fixative solution to 20–22 °C in the water bath or on ice. The final composition of the fixative is 3.7% paraformaldehyde, 50 mM PIPES, pH 6.8, 5 mM EGTA, 2 mM $MgSO_4$, 0.5% (v/v) Triton X-100.
8. Sieve 10 mL of embryogenic cell culture through a 40- to 60-μM mesh strainer (e.g., BD Falcon Cell Strainers from Becton Dickinson, cat. no. 352340), put the cell strainer into a well of a six-well tissue culture plate (e.g., cat. no. 831839, Sarstedt) and add 5 mL of the fixative solution.
9. Fix the sample for 40 min at room temperature.
10. Wash three times, 10 min each, in PBS. If fixed embryos are to be stored, replace PBS with a solution of 0.02% (w/v) sodium azide in PBS and keep at 4 °C. In this way, fixed cells can be stored for a week.

3.7. Attachment of Embryos to the Cover Slips

1. Place about 200 μL of the embryos on a poly-L-lysine-coated cover slip (the embryo slurry must be concentrated enough to cover the whole cover slip) and incubate in a humid chamber for 40 min. Prepare at least two cover slips per staining as the efficiency of the adhesion may vary (*see* **Note 16**).
2. Carefully remove the buffer from the cover slip by touching its corner with a piece of tissue trying not to disturb the embryos. After this, keep the cover slips in the humid chamber for a further 10 min.
3. Place each cover slip into a well of a six-well tissue culture plate containing 3 mL of PBS, wash for 5 min and change the PBS buffer. Use the tissue culture plate to carry out the washes at all subsequent stages.
4. Check the cover slips under an inverted microscope and choose appropriate slides for subsequent stages. The cover slips with cells can be stored in PBS containing 0.02% (w/v) of sodium azide at 4 °C for upto a week.

3.8. Incubation with the Antibodies and Mounting

1. Incubate the cover slips with PBS/BSA/SA buffer for 30 min.
2. Dilute primary antibody in PBS/BSA/SA buffer. For example for the DM1A antibody, use a 1:200 dilution (*see* **Note 17**).
3. Place cover slip cell side up on a grid or slice of parafilm in the humid chamber, then carefully apply 70–100 μL of the primary antibody solution and incubate for 3 h.
4. Wash the cover slips in the six-well tissue culture dishes three times, 20 min each, in 3 mL of PBS.
5. Dilute secondary antibodies as recommended by the manufacturer, apply in the same way as the primary antibody described in **step 3** and incubate for 3 h.
6. Wash cover slips two times, 5 min each, in PBS.
7. Stain DNA in the cells with 3 mL of DAPI/PBS/SA buffer for 10 min.

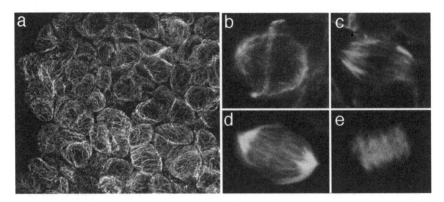

Fig. 2. Microtubule arrays in the embryonal mass cells. (**A**) Interphase cortical microtubules; (**B**) pre-prophase band; (**C**) metaphase spindle; (**D**) anaphase spindle; (**E**) phragmoplast. The images were recorded using a Zeiss 510 laser confocal microscope. Each image represents a combination of three optical sections.

8. Take a clean microscope slide and put a drop (50 µL) of Vectashield mounting medium onto the slide. Aspirate cover slips from the corner and place them cell side down on the drop of mounting medium. Avoid air bubbles being trapped under the cover slip (two cover slips can be mounted per slide). Gently blot the mounted samples on the paper towel or tissue to remove excess mounting medium. Seal the edges of the cover slip with nail varnish. When the nail varnish is dry, carefully rinse the surface with water and dry using a paper towel.

9. Samples can now be stored at 4 °C in the dark. The quality of the staining will gradually deteriorate and the images should be taken as soon as possible (*see* **Note 18**). A typical microtubule organization observed in embryonal mass cells is shown in **Fig. 2**.

3.9. Staining Actin Filaments

1. Incubate cells with rhodamine-phalloidin solution in PBS buffer for 30 min.
2. Mount and observe immediately (*see* **Note 19**).

4. Notes

1. A pre-made acrylamide stock can be purchased from a variety of suppliers; for example, cat. no. EC-890, Protogel from National Diagnostics.
2. Ammonium persulphate deteriorates over time resulting in poor acrylamide polymerization. Use a fresh batch of chemical, aliquot the stock solution and keep at –20 °C. Use each aliquot within a day and dispose leftovers.
3. To prepare PEM buffer, first put PIPES powder into a beaker and add water to two-thirds of the final volume. Stir on the magnetic stirrer and measure the

pH at the same time. Adjust to pH 6.8 using 10 M NaOH; the solution will gradually clear as the pH rises. Add EGTA and MgSO$_4$, check the pH and adjust if necessary. If the pH is above 7.0, do not add acid to reduce the pH as addition of acid will introduce salt into the buffer and will affect the efficiency of the fixation, adjust pH by adding PIPES powder. The fixation will be efficient over the pH range between 6.8 and 7.0.

4. The final concentration of acrylamide/bis-acrylamide in the gel mixture depends on the molecular mass of the protein to be detected. Use 7.5% (w/v) for 50- to 200-kDa proteins, 12.5% (w/v) for 20- to 50-kDa proteins, 15% (w/v) for 10- to 15-kDa proteins.

5. The staining of proteins on the transfer membrane with the weak solution of amido black is reversible and will be hardly visible or will disappear by the end of the probing.

6. If using non-concentrated monoclonal antibodies from the spent tissue culture medium, use dilutions 1:1, 1:5 and 1:10.

7. If a limited amount of antibody is available, the strips can be arranged protein side up on a peace of parafilm placed in a humid chamber and overlaid with the antibody solution. In this case, 70–100 µL of the antibody solution is sufficient to cover the strip.

8. If the signal is to be detected by a digital imaging system, perform these steps in the lab, making sure that the imaging system is ready. Acquire exposures every minute.

9. Some proteins are unstable and can degrade rapidly even when using the boiling SDS–PAGE buffer described resulting in a detection of smaller protein bands on the western blots. In this case, try a protein extraction using the trichloroacetic acid method.

10. If the developed western blots show a strong signal together with high background, increase the concentration of sodium chloride in the TBST buffer upto 0.3–1 M.

11. From this stage on, it is imperative to keep the cover slips clean. Wash fine tweezers with detergent and deionized water, then rinse in 70% ethanol and wipe with clean paper towel avoiding touching the tips. Always keep the tweezers vertical during the washing stages. Wear gloves and wash them with detergent and distilled water before handling the cover slips.

12. The cover slips coated with poly-L-lysine are stable for several months if stored in a dry cool environment.

13. Paraformaldehyde deteriorates when in contact with oxygen and is shipped under argon. Store paraformaldehyde at 4 °C for no longer than 1 year. An old batch of paraformaldehyde will not fix cells efficiently. Some protocols recommend adding glutaraldehyde to the fixative mix. Glutaraldehyde improves the efficiency of fixation, but generates a background that can cause problems when an epifluorescence microscope is used to record the images. The background is not a big problem when the imaging is done using a confocal

microscope. The background can be reduced by incubating the fixed samples with a solution of 10 mM NH_4Cl or NH_4Br in PBS for 20 min.

14. The procedure for preparation of fixative described here is for 20 mL final volume, however, it can be scaled up as required. Always prepare fresh solution.

15. During this step, paraformaldehyde forms the formaldehyde solution. Formaldehyde is toxic and volatile (especially in a warm solution). Handle the fixative solution and carry out all the fixation steps in a fume cupboard.

16. The adhesion of embryos to poly-L-lysine cover slips decreases after treatment with dimethylsulphoxide [DMSO; even concentration of 0.2% (v/v) can have an effect], for example, when a drug stock solution made in DMSO is applied to the embryo culture. Use other solvents to prepare the drug stock solution. If DMSO cannot be avoided, samples can be probed with antibodies in a 'stainer'. A stainer can be made from a 2-mL microcentrifuge tube, a 1-mL pipette tip and 50-µM mesh. First, make a small tube by cutting a 2-mL microcentrifuge tube 1 cm from the bottom, throw away the top half. Cut the tip off of a blue 1-mL pipette tip, so the diameter of the opening of the tip is about 4 mm. Trim the blue tip to 2 cm in length. Melt the 4-mm opening on the Bunsen burner and stick it on to a peace of nylon mesh. Carefully trim the mesh with sharp scissors or razor blade and check if the staining chamber fits into the tube.

17. Several primary antibodies can be mixed together if required, provided they have been raised in different host species and provided that adequate controls are done to ensure that the secondary antibodies do not cross-react between the different species. Secondary antibodies that have minimal crossreactivity with immunoglobulins from different species are available commercially.

18. Sometimes an epitope recognized by an antibody can be hidden within the tertiary structure of the protein or within a protein complex and is therefore inaccessible to the antibody. As a result, a good signal on a western blot may be accompanied by a negative result in immunofluorescence. If this does happen, then treating the fixed cells with methanol and acetone may help. Incubate formaldehyde fixed cells with −20 °C methanol for 10 min followed by acetone at −20 °C for 5 min. Both methanol and acetone have to be chilled to −20 °C before the treatment. After this, cells are allowed to re-hydrate in PBS for 30 min, and washed two times, 5 min each, in PBS before incubation with PBS + BSA solution. The treatment can be done either while the cells are attached to the cover slips or in the stainers.

19. The interaction of phalloidin with fixed actin is not stable, and samples have to be imaged immediately after mounting. Visualization of actin with phalloidin-based probes may not work if the samples were treated with methanol/acetone.

References

1. Hyams JS, Lloyd CW, eds. Microtubules. New-York: Wile-Liss, 1994.
2. Hussey PJ, ed. The Plant Cytoskeleton in Cell Differentiation and Development. Oxford: Blackwell Publishing, 2004.

3. Goddard RH, Wick SM, Silflow CD, Snustad PD. Microtubule components of the plant cell cytoskeleton. Plant Physiol 1994;104:1–6.
4. Paredez AR, Somerville CR, Ehrhardt DW. Visualization of cellulose synthase demonstrates functional association with microtubules. Science 2006;312:1491–1495.
5. Hussey PJ, Ketelaar T, Deeks MJ. Control of the actin cytoskeleton in plant cell growth. Annu Rev Plant Biol 2006;57:109–125.
6. Ketelaar T, Anthony RG, Hussey PJ. Green fluorescent protein-mTalin causes defects in actin organization and cell expansion in Arabidopsis and inhibits actin depolymerizing factor's actin depolymerizing activity in vitro. Plant Physiol 2004;136:3990–3998.
7. Smertenko AP, Bozhkov PV, Filonova LH, von Arnold S, Hussey PJ. Reorganisation of the cytoskeleton during developmental programmed cell death in *Picea abies* embryos. Plant J 2003;33:813–824.
8. Bozhkov PV, von Arnold S. Polyethylene glycol promotes maturation but inhibits further development of *Picea abies* somatic embryos. Physiol Plant 1998;104:211–224.
9. von Arnold S, Eriksson T. A revised medium for growth of pea mesophyll protoplasts. Physiol Plant 1977;39:257–260.
10. Laemmli UK. Cleavage of structural proteins during assembly of head of bacteriophage-T4. Nature 1970;227:680–685.

14

Detection of Programmed Cell Death in Plant Embryos

Lada H. Filonova, María F. Suárez, and Peter V. Bozhkov

Summary

Programmed cell death (PCD) is an integral part of embryogenesis. In plant embryos, PCD functions during terminal differentiation and elimination of the temporary organ, suspensor, as well as during establishment of provascular system. Embryo abortion is another example of embryonic PCD activated at pathological situations and in polyembryonic seeds. Recent studies identified the sequence of cytological events leading to cellular self-destruction in plant embryos. As in most if not all the developmental cell deaths in plants, embryonic PCD is hallmarked by autophagic degradation of the cytoplasm and nuclear disassembly that includes breakdown of the nuclear envelope and DNA fragmentation. The optimized setup of terminal deoxynucleotidyl transferase-mediated dUTP nick end labeling (TUNEL) allows the routine in situ analysis of nuclear DNA fragmentation in plant embryos. This chapter provides step-by-step procedure of how to process embryos for TUNEL and how to combine TUNEL with immunolocalization of the protein of interest.

Key Words: Chromatin; DNA fragmentation; immunolocalization; plant embryogenesis; programmed cell death; TUNEL.

1. Introduction

Breakdown of nuclear envelope and DNA fragmentation are the major events in eukaryotic programmed cell death (PCD) executed through activation of specific proteases and nucleases. The process of nuclear envelope disassembly in plants and animals is morphologically similar and encompasses dismantling of nuclear pore complex and lobing of the nuclear surface leading to nuclear segmentation. Likewise, the pattern of nuclear DNA fragmentation during plant PCD resembles apoptotic DNA fragmentation yielding long (50 kbp),

From: *Methods in Molecular Biology, vol. 427: Plant Embryogenesis*
Edited by: M. F. Suárez and P. V. Bozhkov © Humana Press, Totowa, NJ

chromatin loop-sized fragments and internucleosomal (180 bp and multiples thereof) fragments *(1,2)*. Release of endonuclease ZEN1 from collapsed lytic vacuoles and nuclear accumulation of a type-II metacaspase protease are cellular mechanisms causing nuclear degradation at plant PCD *(3,4)*.

Nuclear DNA fragmentation represents a good biochemical marker for detecting PCD in diverse eukaryotic systems. There is an ever increasing number of methods developed for biochemical and cytochemical analyses of DNA fragmentation *(5)*. Where the aim is to localize PCD during normal pattern formation or aberrant tissue/organ patterning in wild-type embryos or embryo-specific mutants, in situ analysis of DNA fragmentation by terminal deoxynucleotidyl transferase-mediated dUTP nick end labeling, or TUNEL, is apparently the most informative method. This method uses the enzyme terminal deoxynucleotidyl transferase to add fluorescently or biotin-labeled deoxynucleotide triphosphates (dUTP), in a template-independent manner, to the 3′ OH ends of either single- or double-stranded DNA *(6)*. When embryos are stained with TUNEL and counterstained with a nuclear dye [usually 4,6-diamidino-2-phenylindole (DAPI)] for microscopy analysis, one can simultaneously assess cellular morphology, shape of the nuclei and DNA fragmentation, making this a powerful tool for PCD detection.

Using well-established model system of Norway spruce somatic embryogenesis *(2)*, TUNEL analysis was used to monitor DNA fragmentation over the whole pathway of embryonic pattern formation and embryo maturation. DNA fragmentation was found to be restricted to the embryo suspensor during apical–basal pattern formation, to provascular tissue at the stages of active histogenesis and also to cotyledons and root cap in mature embryos *(7)*. These examples of morphogenetic or histogenetic cell deaths are indispensable for successful embryo development. If embryogenesis goes off the normal pathway or blocked, the whole embryo becomes subject to autodestruction, resulting in the massive TUNEL throughout the whole embryo *(8,9)*.

Molecular mechanisms of plant PCD are poorly understood. Most of the metazoan pro- and anti-cell death proteins are absent in plant genomes. Therefore, genomic and proteomic approaches are currently undertaken to identify molecular components of plant PCD *(10,11)*. Unraveling cellular and biochemical functions of these components would require intracellular localization of the protein in living cells and at the successive stages of PCD in the developmental context. Combining immunocytochemistry with TUNEL is a straightforward approach, which has previously been used for studying the integrity of microtubules and intracellular translocation of metacaspase during PCD in spruce embryos *(4,9)*. Here, we describe the protocol for simultaneous localization of nuclear DNA fragmentation and protein of interest in plant embryos.

2. Materials

2.1. Preparation of Whole-Mount Samples

1. Immature or mature seeds or somatic embryos.
2. Dumont tweezers, scalpel handles and blades and dissecting needles from Agar Scientific.
3. Cell strainers, 70-μm pores (BD Biosciences). We recommend to cut off extended lips by scissors.
4. Vacuum desiccator and pump (Fisher).
5. Polysine microscope slides (Menzel-Glaser).
6. Phosphate-buffered saline (PBS): Prepare 10× stock with 1.37 M NaCl, 27 mM KCl, 100 mM Na_2HPO_4, 18 mM KH_2PO_4 (adjust to pH 7.4 with HCl if necessary) and autoclave before storage at room temperature. Prepare working solution by dilution of one part with nine parts water.
7. PEM buffer: 50 mM 1,4-piperazinediethanesulfonic acid (from Sigma), 5 mM ethylene–bis(oxyethylenenitrilo)tetraacetic acid (EGTA; from Sigma), and 2 mM $MgSO_4$ (adjust to pH 6.8 with HCl or NaOH if necessary). Store at 4 °C.
8. Fixative solution: 3.7% (w/v) paraformaldehyde (Fisher) in PEM buffer, with the addition of 0.5% (v/v) Triton X-100. The solution may need to be heated to 70 °C with addition of 0.1 M NaOH (50 μL per 20 mL solution) to dissolve. Then cool to room temperature for use. Always use freshly prepared solution for each experiment.
9. Hydrolytic enzyme cocktail: 1% (w/v) Macerozyme R-10 (Sigma) and 0.2% (w/v) Pectolyase (Sigma) in 15 mM 4-morpholineethanesulfonic acid (adjust to pH 5.0), supplemented with 5 mM EGTA and 0.4 M mannitol (Sigma). Filter through 0.45-μm syringe filter (VWR Int.) and store in 5 mL aliquots at –20 °C. Before use, add proteinase inhibitors (per each 1 mL): 10 μL of 100 mM phenylmethane-sulfonyl fluoride, 1 μL of 10 mg/mL leupeptine and 1 μL of 1 mg/mL pepstatine A (all from Sigma).

2.2. TUNEL

1. Proteinase K: To prepare stock solution, dissolve nuclease free Proteinase K (Sigma) at 10 g/mL in DNase free water. Store at –20 °C. Prepare working solution by dilution of the stock with 0.1 M Tris-HCl (pH 7.5) to a final concentration of 20 μg/mL.
2. Blocking solution: 3% (w/v) bovine serum albumin (Sigma) and 20% (v/v) normal bovine serum (GeneTex Inc.) in 0.1 M Tris–HCl (pH 7.5). Filter solution through a 0.45-μm syringe filter (VWR Int.) and store at 4 °C for 1 week.
3. TUNEL: In Situ Cell Death Detection Kit, Fluorescein (11 684 795 001, Roche Diagnostics) or In Situ Cell Death Detection Kit, TMR red (12 156 792 001, Boehringer Mannheim). Store in aliquots at –20 °C. Once thawed, store at 4 °C for 1–4 days. To prepare TUNEL reaction mixture, add one volume of Enzyme Solution (terminal transferase) to nine volumes of Label Solution (nucleotide mix). Keep on ice.

2.3. Antibody Staining

1. Primary antibody. Follow appropriate storage recommendations and dilutions (see product information sheet).
2. FITC- or TRITC-conjugated secondary antibody. Follow appropriate storage recommendations and dilutions (see product information sheet).

2.4. DAPI Staining

1. S-buffer: 0.25 M sucrose, 1 mM ethylenediamine tetraacetic acid, 0.6 mM spermidine, 0.05% (v/v) mercaptoethanol in 10 mM Tris–HCl, (pH 7.6). Store at 4 °C.
2. DAPI stock solution: dissolve 4 mg DAPI (Sigma) in 1 mL S-buffer. Store in aliquots at –20 °C. Stock can be thawed several times without loosing its staining efficiency.
3. DAPI solution. Add 2 µL of DAPI stock to 10 mL PBS. Solution can be stored at 4 °C in the dark for up to 2 months.

2.5. Mounting Samples and Microscopy

1. Cover slips (24 × 60 mm) from Menzel-Glaser.
2. FluorSave™ Reagent (Calbiochem).
3. Nail varnish.

3. Methods

The method described here can be used for embryos from any plant species, regardless of whether embryos are large (e.g., in conifers) or minute (e.g., in *Arabidopsis*). Likewise the method works equally well both with somatic and zygotic embryos.

3.1. Preparation of Whole-Mount Samples

Tissue sectioning can sometimes affect nuclear integrity leading to appearance of false-positive TUNEL signals *(12)*. Therefore, we recommend working with the whole-mount embryos, which also saves much time as compared to sectioned embryos. Because harsh physical manipulations with the ovules can wound embryonic tissues and produce TUNEL reactivity, it is important to fix the whole ovules before embryo isolation.

1. Transfer ovules (*see* **Note 1**) or somatic embryos into cell strainers placed inside small glass beakers.
2. Put beakers into a vacuum desiccator.
3. Add fixative solution to the samples. Close the desiccator and apply a vacuum to the samples until the fixative solution starts to bubble.

4. Incubate the samples in the fixative solution for 1 h at room temperature.
5. Discard the fixative solution into a paraformaldehyde waste container and wash the samples twice for 5 min each with PEM buffer.
6. Place the samples in drops of PEM buffer on Polysine slides. Large zygotic embryos should now be dissected from the ovules under stereomicroscope using thin tweezers and dissecting needles or surgical blades. Discard the ovules.
7. Carefully remove the buffer from the slides and let the samples to air-dry for 1 h at room temperature.
8. For small embryos, proceed to **Subheading 3.2.** For large embryos, incubate the slides with hydrolytic enzyme cocktail for 15 min at room temperature to permeabilize the tissues, and then wash the slides twice for 5 min each in PBS.

3.2. TUNEL

1. If TUNEL is not followed by antibody staining, treat the samples with Proteinase K solution for 15 min at room temperature, followed by two washes, 10 min each, with distilled water. Otherwise, proceed directly to step 2.
2. Incubate the samples in blocking solution for 30 min at room temperature.
3. Discard the blocking solution and wash the samples twice for 15 min each with PBS.
4. Add TUNEL reaction mixture to the samples (*see* **Note 2**). Then put the slides carefully into moisture-saturated chamber above the water (*see* **Note 3**), close the chamber and incubate for 1 h at 37 °C. As a negative control for TUNEL, use a separate slide incubated with the label solution (without terminal transferase).
5. Wash the samples twice, 10 min each, with PBS, and then rapidly twice with distilled water.

3.3. Antibody Staining

1. Incubate TUNEL-stained samples with primary antibody for 2 h at room temperature or at 4 °C overnight. For negative control, proceed directly to step 3.
2. Remove the primary antibody and wash the samples three times for 10 min each with PBS.
3. Incubate the samples with secondary antibody for 1 h at room temperature.
4. Remove the secondary antibody and wash the samples twice for 5 min each with PBS.

3.4. DAPI Staining

1. Incubate the samples with DAPI solution for 10 min at room temperature to stain DNA and identify the nuclei.
2. Wash the samples three times for 5 min each with PBS and then aspirate dry from one corner.

3.5. Mounting Samples and Microscopy

1. Put a drop of Fluorsave onto a cover slip, invert a cover slip and place gently on Polysine slide with the samples. Use nail varnish to seal cover slip. The samples can be viewed immediately that the varnish is dry, or be stored in the dark at 4 °C for up to a month.
2. View the slides under phase contrast microscopy (to locate the embryos and define basal–apical pattern) and then under epifluorescent or confocal microscopy. Excitation at 543 nm induces TUNEL TMR-red or TRITC-conjugated IgG fluorescence (red emission), excitation at 490 nm induces TUNEL Fluorescein or FITC-conjugated IgG fluorescence (green emission), while excitation at 364 nm induces DAPI fluorescence (blue emission). Record images via CCD camera using unified exposure time for all samples to detect differences in signal intensity. Software can be used to overlay the phase contrast and fluorescence images.

4. Notes

1. If the seeds are large (e.g., in conifers), the testa should be removed. Minute seeds with soft testa (e.g., in *Arabidopsis*) are used as a whole.
2. For economy, only 100 μL of the TUNEL reaction mixture per slide needs to be used.
3. We use plastic slide storage box (Electron Microscopy Sciences) with a thin layer of water.

Acknowledgments

The cell death research in our group is supported by the Carl Tryggers Foundation, the Swedish Foundation for International Cooperation in Research and Higher Education, the Spanish Ministry of Education and Science, the Wenner-Gren Foundation, the Swedish Research Council (VR) and the Swedish Research Council for Environment, Agricultural Sciences and Environmental Planning (Formas).

References

1. Earnshaw WC. Nuclear changes in apoptosis. Curr Opin Cell Biol 1995;7:337–343.
2. Bozhkov PV, Filonova LH, Suárez MF. Programmed cell death in plant embryogenesis. Curr Top Dev Biol 2005;67:135–179.
3. Ito J, Fukuda H. ZEN1 is a key enzyme in the degradation of nuclear DNA during programmed cell death of tracheary elements. Plant Cell 2002;14:3201–3211.
4. Bozhkov PV, Suárez MF, Filonova LH, Daniel G, Zamyatni, AA Jr, Rodriguez-Nieto S, Zhivotovsky B, Smertenko A. Cysteine protease mcII-Pa executes programmed cell death during plant embryogenesis. Proc Natl Acad Sci USA 2005;102:14463–14468.

5. Reed JC, ed. Apoptosis. Methods in Enzymology, vol. 322. San Diego: Academic Press, 2000.

6. Gavrieli Y, Sherman Y, Ben-Sasson SA. Identification of programmed cell death in situ via specific labeling of nuclear DNA fragmentation. J Cell Biol 1992;119: 493–501.

7. Filonova LH, Bozhkov PV, Brukhin VB, Daniel G, Zhivotovsky B, von Arnold S. Two waves of programmed cell death occur during formation and development of somatic embryos in the gymnosperm, Norway spruce. J Cell Sci 2000;113:4399–4411.

8. Filonova LH, von Arnold S, Daniel G, Bozhkov PV. Programmed cell death eliminates all but one embryo in a polyembryonic plant seed. Cell Death Differ 2002;9:1057–1062.

9. Smertenko AP, Bozhkov PV, Filonova LH, von Arnold S, Hussey PJ. Reorganisation of the cytoskeleton during developmental programmed cell death in *Picea abies* embryos. Plant J 2003;33:813–824.

10. Demura T, Tashiro G, Horiguchi G, Kishimoto N, Kubo M, Matsuoka N. Visualization by comprehensive microarray analysis of gene expression programs during transdifferentiation of mesophyll cells into xylem cells. Proc Natl Acad Sci USA 2002;9:15794–15799.

11. Taylor NL, Heazlewood JL, Day DA, Millar AH. Differential impact of environmental stresses on the pea mitochondrial proteome. Mol Cell Proteomics 2005;4:1122–1133.

12. Sloop GD, Roa JC, Delgado AG, Balart JT, Hines MO III, Hill JM. Histologic sectioning produces TUNEL reactivity. A potential cause of false-positive staining. Arch Pathol Lab Med 1999;6:529–532.

Subject Index

A

Abscisic acid (ABA), 33, 41
Actin filaments, 35–36, 158, 168
Agarose layer, enlarged microspores
    immobilization, 85–87; *see also* Barley
    microspore embryogenesis
Agrobacterium tumefaciens, 91–93, 95
    strain GV3101, 131
Amplified RNA, 114
APETALA2-domain transcription factors, 9
Arabidopsis Biological Resource Center, 122
Arabidopsis embryogenesis, 3
    apical–basal patterning, 6–10
    asymmetric division, of zygote, 5–6
    auxin visualization (*see* Auxin visualization,
        in embryogenesis)
    embryo-specific mutants isolation in, 101
        cosegregation analysis, 106
        emb mutants, 102
        genetic screen, for putative embryo mutants,
          103–105
        reciprocal crosses, 105
        whole-mount clearing method, phenotypic
          analysis by, 105–106
    heart-stage embryo, 7
    "*in-planta*" transformation method, 92
    LCM method (*see* Laser-capture microdissection
        method)
    morphological development, 4–5
    plant transformation
        Agrobacterium culture and preparation, 93, 95
        Agrobacterium-mediated gene transfer in, 91
        plant cultivation, 92, 94–95
        transformants screening, 97
        vacuum infiltration and floral dip
          transformation, 95–97
    promoter trapping system and (*see* Promoter
        trapping system, embryogenesis and)
    radial pattern formation, 10–11
    in vitro culture (IVC), 71

culture plates preparation, 72–73
expressing fluorescent proteins of, microscopic
    analysis, 74
histochemical methods, 74
microscopic morphological embryo analysis,
    73–74
transfer to AM medium for long-term
    culture, 73
Arabidopsis thaliana (L.), 3
ARABIDOPSIS THALIANA MERISTEM LAYER1
    (*ATML1*) genes, 5, 10
AREs, *see* Auxin responsive elements
ARF, *see* Auxin response factor
aRNA, *see* Amplified RNA
Aux/indole-3-acetic acids (Aux/IAAs), 138–139
AUXIN RESISTANT6 (AXR6), 8–9
Auxin response factor, 8, 138
Auxin responsive elements, 138
Auxin visualization, in embryogenesis, 137
    auxin responsive transgenes and, 138
    direct, by immunolocalization, 140
    with fluorescent auxin responsive reportergenes,
        139, 140–141
    GUC-based reporter genes and, 139, 141–142
    immunocytochemical, 139

B

Barley microspore embryogenesis, 77
    anther pre-treatment, 80–82
    cell tracking systems and, 78
    culture media and stock solutions, 79–80
    donor plants, growth, 80
    microspore culture
        cell tracking, 84–87
        liquid, 83–84
        solid, 84
    microspore isolation, 82–83
    mid-late to late (ML-L) uninucleate stage of
        microspore, 78
    star-like microspores stage, 78

Basic Local Alignment Search Tool (BLAST), 125, 130
BODENLOS (BDL), 8–9
BRASSINOSTEROID-INSENSITIVE1, 8

C
Cauliflower mosaic virus 35S promoter, 132
cDNA libraries, 53
Cell tracking systems, 78; *see also* Barley microspore embryogenesis
enlarged microspores
immobilization, in agarose layer, 85–87
purification, by sucrose gradient, 84–85
clavata1/2/3 (clv1/2/3) mutants, 7
CUP-SHAPED COTYLEDON (CUC) genes, 7
CYTOKININ RESPONSE1, 11

D
DAPI, *see* 4,6-Diamidino-2-phenylindole
Deoxynucleotide triphosphates, 174
4,6-Diamidino-2-phenylindole, 174, 177
Dot-blot analysis, 23–24
DR5rev, auxininducible promoter, 5
DR5rev::GFP, 139
dUTP, *see* deoxynucleotide triphosphates

E
Embryodefective mutants *(emb),* 102

F
fackel (fk) mutants, 8
F-actin, *see* Actin filaments
Fluorochrome, 158

G
Genetic screen, for putative embryo mutants; *see also Arabidopsis*
Kanamycin selection marker, segregation ratio analysis of, 103–104
for reduced seed set, 104–105
GFP, *see* Green fluorescent protein
β-Glucuronidase, 74, 121, 138
based reporter genes, auxin response visualization by, 139, 141
expression patterns, screening for, 122–124
GUS-stained siliques sectioning, 127
GUS staining, 126
seed coat clarification, 126
silique dissection, 125
tagged gene expression and, 130–131
GRAS-type transcription factors, 11
Green fluorescent protein, 138

GURKE gene, 6
GUS, *see* β-Glucuronidase
GUS reporter gene, 131

H
HBK1, HBK2 and *HBK3, KNOX (KNOTTED1*-like homeobox) family, 36
HD-GL2, 36
HOBBIT (HBT) gene, 10
Hordeum vulgare L., *see* Barley
HotStarTaq Mix, 62

I
Immunocytochemical techniques, *in vitro* fertilization, 55–56, 62–64
In vitro culture, of *Arabidopsis* embryos, 71; *see also Arabidopsis*
culture plates preparation, 72–73
expressing fluorescent proteins of, microscopic analysis, 74
histochemical methods, 74
microscopic morphological embryo analysis, 73–74
transfer to AM medium for long-term culture, 73
In vitro fertilization, with isolated higher plant gametes, 51
fusiogenic media for, 52
fusion of gametes, 54, 57–58
genes, identification and expression, 55
PCR-based approach, 61–62
immunocytochemical techniques, 55–56, 62–64
isolation and transfer of gametes, 53–54, 56–57
proteins identification, 55, 59–61
single-cell micromanipulation techniques, 53
zygotes and central cells culture, embryogenesis and regeneration, 55, 58–59
IVC, *see In vitro* culture

K
Kanamycin (Kan) selection marker, 102–104

L
Laser-capture microdissection method, 111, 115
cryosections of globular-stage *Arabidopsis* embryo, 112–113
RNA amplification, 116–117
RNA extraction, 116
stages of, 114
LCM method, *see* Laser-capture microdissection method
LC-MS/MS analysis, 59, 66

M
Maize embryogenesis, 17
  growing maize, in greenhouse, 18, 20
  immature embryo culture, 18, 20–21
  in situ hybridization, 19
    DIG detection, 26–27
    dot-blot analysis, 23–24
    fixation, embedding and sectioning, 24–25
    pre-hybridization treatments, 25–26
    probe labeling, 22–23
  somatic, 18–19, 21–22
Microtubule arrays, 158, 168
Microtubule-associated proteins, 157–158
Millicell-CM inserts, 62, 66
Mitogen-activated protein (MAP) kinase, 6
Monopteros (mp) mutant, 8

N
NAC transcription factors, 7
Norway spruce embryogenesis
  cryopreservation, 38
    dimethyl-sulfoxide cryoprotective
      treatment, 42
    freezing and thawing, 42–43
    sorbitol treatment, 42
  gene transfer, 38, 43–44
  somatic
    developmental pathway of, 35
    system of, 36–42
    time-lapse tracking technique, 33–34
  zygotic, 31–32
NptII gene, 122

P
PaHB1 (Picea abies Homeobox1), 36
PALM® MicroLaser system, 112
PCD, *see* Programmed cell death
pDeltaGUS promoter trap system, T-DNA and, 122
PEG, *see* Polyethylene glycol
PEMs, *see* Proembryogenic masses
pGEM-T Easy vector, 62
Picea abies, see Norway spruce
PIN FORMED7 (PIN7) gene, 5
PINOID, serine/threonine kinase, 8
PixCell II system, 114
Plasmodesmata, 145
  probe-movement assays, 147–149
  size exclusion limit (SEL), 145–146
    embryo-defective mutants isolation and,
      150–152
PLETHORA (PLT) genes, 9–10
Polyethylene glycol, 45

Poly-L-lysine, 166
Proembryogenic masses, 31, 33, 36, 39
Programmed cell death detection, in plant
    embryos, 173
  antibody staining, 176–177
  DAPI staining, 176–177
  mounting samples and microscopy, 178
  nuclear DNA fragmentation and, 174
  TUNEL analysis, 174–175, 177
  whole-mount samples preparation, 175–177
Promoter trapping system, embryogenesis and
  DNA sequences flanking insertion,
      amplification, 124
    DNA isolation, 128
    TAIL-PCR, 128–129
  GUS expression patterns, screening for, 122–124
    GUS-stained siliques, sectioning, 127
    GUS staining, 126
    seed coat clarification, 126
    silique dissection, 125
  TAIL-PCR products sequencing, 124, 129
    sequence analysis, 130
    validation of, 130
Proteins immunolocalization, in somatic
    embryos, 157
  antibodies, incubation with, 168–169
  fixation and permeabilization, of embryos, 162,
      166–167
  immunohistochemistry and, 158
  indirect immunofluorescence microscopy, 158
  microtubule arrays, 168
  poly-L-lysine-coated cover slips preparation,
      161, 166
  primary antibody, source of, 159
  protein extraction, 160
  SDS-PAGE, 160–161, 163–164
  staining actin filaments, 168
  tissue culture, 160, 162
  Western blotting, 161, 164–165
PROTODERMAL FACTOR2 (PDF2), 10
PtNIP1;1 gene, 36

R
Radish *(Raphanus sativus)*, 114
Reverse transcription-polymerase chain reaction,
    53, 130
RNA amplification, 116–117
RNA extraction, 116
RNA™ Nanoprep kit, 116
RT-PCR methods, *see* Reverse
    transcription-polymerase chain reaction

S
SCARECROW (SCR) genes, 9–11
SDS-PAGE, *see* Sodium dodecyl sulphate
    poly-acrylamide gel electrophoresis
SHOOT MERISTEMLESS (STM), 7
SHORTROOT (SHR) genes, 10–11
Single-cell micromanipulation techniques, 53
Sodium dodecyl sulphate poly-acrylamide gel
    electrophoresis, 59–61, 66, 160–161, 163–164
Somatic embryogenesis
    maize, 18–19, 21–22
    in Norway spruce
        developmental pathway of, 35
        system of, 36–42
        time-lapse tracking technique, 33–34
    proteins immunolocalization and, 157
        antibodies, incubation with, 168–169
        fixation and permeabilization, of embryos,
            162, 166–167
        immunohistochemistry and, 158
        indirect immunofluorescence microscopy, 158
        microtubule arrays, 168
        poly-L-lysine-coated cover slips preparation,
            161, 166
        primary antibody, source of, 159
        protein extraction, 160
        SDS-PAGE, 160–161, 163–164
        staining actin filaments, 168
        tissue culture, 160, 162
        Western blotting, 161, 164–165
Super SMART™ PCR cDNA Synthesis Kit, 62

T
TAIL-PCR, *see* Thermal asymmetric interlaced
    polymerase chain reaction
T-DNA, *see* Transfer-DNA
Terminal deoxynucleotidyl transferase-mediated
    dUTP nick end labeling, 174–175, 177
N,N,N',N'-Tetramethylethylenediamide
    (TEMED), 163

Thermal asymmetric interlaced polymerase chain
    reaction, 124, 128
    products, sequencing of, 129–130
TOPLESS gene, 6
Transfer-DNA, 91–92, 101, 106, 131
    insertion, DNA sequences amplification and,
        128–129
    pDeltaGUS promoter trap system and, 122
T7 RNA polymerase-mediated amplification
    technique, 114
Ttime-lapse tracking technique, of somatic embryo
    formation, 33–34
TUNEL, *see* Terminal deoxynucleotidyl
    transferase-mediated dUTP nick end labeling

U
uidA gene, 36

V
Vacuum infiltration and floral dip transformation,
    of *Arabidopsis,* 95–97; *see also Arabidopsis*
Viviparous 1 gene *(Pavp1),* 36

W
Western blotting, 161, 164–165
Whole-mount clearing method, 105–106
wooden leg (wol) mutation, 11
WUSCHEL-RELATED HOMEOBOX (WOX)
    genes, 5
WUSCHEL (WUS) gene, 7

Y
Yeast (YOL34), 159
YODA gene, 6

Z
Zea mays L., *see* Maize
Zygotic spruce embryogenesis, 31–32